From Quantum Mechanics to Chronomorphy

New Thinking of Time

Peter Paul Lehleiter

2025

Chronomorph Verlag
Morellstraße 19, 86159 Augsburg, Germany

E-mail: Peter.Paul.Lehleiter@web.de

Print: see last page.

ISBN: 978-3-9826647-2-9

Preface

The impulse came from David Bohm. He once wrote that time had not been properly incorporated into quantum mechanics. For me, this was like a spark: if true, it would not be a marginal issue but a fundamental flaw in the self-understanding of physics itself.

I had always been searching for a topic of real significance—something that would not merely refine technical details but would touch the very foundations. The unification of the forces, it seemed to me, had already advanced so far that little remained to be gained. But the fact that time and events themselves remained vague at the very core of physics was electrifying.

From the very beginning, I was convinced: the first steps in the history of thought had already pointed in the qualitatively right direction. The Greeks may not have had modern mathematics, but they saw the cosmos as a unity of time, motion, and meaning. Anyone who wishes to reflect on time must understand these origins before developing new models.

Therefore, in search of a mathematical model of time, I first immersed myself in ancient thought and its treatment of temporal concepts—as a foundation for a new perspective on physics and quantum mechanics. After long periods of reflection, research, practical experiments, and statistical analyses, the present book represents a first attempt at gaining a unified view of physics from the perspective of a new concept of time.

As this first edition is intended for a limited group of reviewers and interested readers, layout and typography have been kept deliberately simple to maintain full consistency with the original manuscripts.

Abstract

This book addresses the problem of time in physics from two complementary angles.

The first part develops a new framework for quantum mechanics, treating time as a real, event-related quantity rather than a mere parameter and interpreting probabilities not as endpoints but as an interface between global structures and local events. This shift opens the possibility of uncovering correlations across scales and establishes structural parallels with relativity theory.

The second part proposes a large-scale statistical approach—a mesoscopic experiment with macroscopic correlations—designed to test these ideas. Using an astronomical time model, it reduces stochasticity in probabilistic data and provides a concrete basis for empirical verification. Three domains are central: atomic spectra, double-slit interference, and the time scale of atomic decay.

The third part extends these considerations to relativity and cosmology. Building on statistical evidence that quantum-like effects appear across scales, it interprets relativity as the flat-space limit of a wave function and explores black holes as oscillatory concentrations of time energy. This framework leads to a cyclical view of the universe, in which coherence, rather than entropy, sets the direction of cosmic evolution.

By unifying quantum mechanics, relativity, and cosmology under a temporal perspective, the work suggests that time is not merely a background parameter but the organizing principle of physical reality itself.

Introduction

The present work addresses a gap in the physical understanding of time across three major domains of physics: quantum mechanics, relativity theory, and cosmology. The argument unfolds in three parts that together form a coherent progression.

Part I develops the conceptual framework. Heisenberg's uncertainty relation, $\Delta x \Delta p \geq \hbar/2$, shows that spatial quantities such as position and momentum are no longer freely accessible. On the quantum level, this prevents any direct intervention without confronting the limits of uncertainty. Within quantum mechanics, the probabilistic description itself leaves no additional degrees of freedom: the wave function is already globally defined and condenses into a local outcome. The so-called "collapse" should therefore not be taken as a final point, but rather as an interface between the established probabilistic formalism and a prospective extension. Probabilities are not an endpoint, but a transition—the translation of global structures into local events. From this perspective, quantum events must allow for further structural resolution: correlations across scales are already implicit in a single probabilistic outcome.

Both quantum mechanics and relativity theory indicate that space and time are inseparable in a fundamental sense. What is required, then, is not another local observable, but a non-local spatio-temporal relation that can be meaningfully connected to the quantum formalism. Our proposal is to shift the focus: from local observables to correlations across different spatio-temporal scales, supported by a generalized and non-scalable conception of time.

Part II provides the empirical counterpart. Having understood in Part I that the probabilistic framework must be extended by testing correlations across scales, we adopt a mesoscopic perspective. Since direct intervention at the quantum level is blocked by the uncertainty relation, the task is to postulate and statistically test the similarity between two temporal scales. As the collapse of the wave function already suggests, the universe is holistic; time may therefore be treated in a coarser sense: no longer a rescalable parameter, but a structure bound to real events. Unlike clock time—where nothing changes if seconds are subdivided into hundredths—the required time concept must be non-scalable. Such a notion can be established by anchoring time to actual events. Moving one scale higher, the most precise and accessible candidates are astronomical cycles. This is consistent with the historical origin of time itself, which was first defined astronomically before being abstracted into mechanical oscillations.

Part III extends the framework to cosmology. The evidence presented in Part II shows with high significance ($p \approx 0.00001$) that we are not dealing with a special case, but with a general phenomenon: quantum-like effects appear across different scales. Quantum mechanics should therefore not be regarded as confined to small distances or low energy differences. Rather, such effects manifest whenever a process is sufficiently rich—almost probabilistic in character. In these cases, non-local correlations arise, indicating that at least two different scales—for example, the microscopic and mesoscopic domains discussed in Part II—act as correlatives.

Generalizing this observation, we may say that such effects can appear on multiple scales. It thus becomes consistent to adopt a modified Planck constant, allowing it to grow arbitrarily large. In the limiting case, special relativity emerges as the flat-space limit of a globally defined wave function. From here, the argument naturally extends to general relativity and cosmology. Applying the temporal model to black holes, we are led to interpret them as oscillatory concentrations of time energy. This perspective suggests that black holes decay much faster than predicted by Hawking radiation.

The circle closes with cosmology. If coherence is the fundamental principle—the central feature of quantum events—then the universe does not end in heat death or a big freeze. Instead, coherence continues to build until it spontaneously condenses, by analogy with a Fourier transformation, into localized energy concentrations. These provide the conditions for a new beginning, a restart of the universe.

Part I

The Temporal Structure of Quantum Events

Toward a New Reading of Quantum Mechanics

Dr. Peter Paul Lehleiter

Independent Researcher and Publisher (self-funded)
Morellstraße 19
86159 Augsburg
Germany

Corresponding author:
Dr. Peter Paul Lehleiter
https://orcid.org/0009-0005-1997-8564
E-mail: Peter.Paul.Lehleiter@web.de

Contents

Abstract

This work develops a framework for quantum mechanics (QM) in which time is treated as a structured, event-related quantity rather than a mere parameter. Quantum states are interpreted as evaluated measures of temporal structure, with the wave function representing a constrained, cyclic flow of such measures.

Starting from the Schrödinger equation, the analysis motivates a generalized Born rule in which *time selection* could reveal phase-dependent subensembles. The standard Born rule then appears as an average over many measurements rather than as a fundamental postulate.

Although the work is primarily conceptual in nature, concrete experimental tests are proposed—via interference patterns, spectral line shapes, and decay statistics—that can examine whether temporal correlations systematically influence quantum outcomes. A mesoscopic model discussed in [4] already provides initial evidence in this regard.

Keywords: Quantum Mechanics, Eventual Time, Phase Coherence, Generalized Born Rule

References

[1] P. P. Lehleiter, *Time Numbers – in Events, Time, and Quantum Mechanics – Supplementary Material*. OSF Project, 2025. Available: https://osf.io/7pgne (see *UsingOSFProject.pdf* in the root folder for project structure and file descriptions).

[2] P. P. Lehleiter, Non Hermitian Quantum Mechanics – in Events, Time, and Quantum Mechanics – Supplementary Material. OSF Project, 2025. Available: https://osf.io/7pgne (see *UsingOSFProject.pdf* in the root folder for project structure and file descriptions).

[3] P. P. Lehleiter, *Wie Zeit sich in Ereignissen manifestiert: Ansätze einer abstrakten Chronomorphie*. Augsburg, Germany: Chronomorph Verlag, 2024.

[4] P. P. Lehleiter, *Events and Time – in Events, Time, and Quantum Mechanics Supplementary Material*. OSF Project, 2025. Available: https://osf.io/7pgne (see *UsingOSFProject.pdf* in the root folder for project structure and file descriptions).

[5] E. Schrödinger, "Quantisierung als Eigenwertproblem. Dritte Mitteilung," *Ann. Phys.*, vol. 80, pp. 437–490, 1926.

[6] E. Schrödinger, "Quantisierung als Eigenwertproblem. Erste Mitteilung," *Ann. Phys.*, vol. 79, pp. 361–376, 1926.

[7] E. Schrödinger, "Quantisierung als Eigenwertproblem (Vierte Mitteilung)," *Ann. Phys.*, vol. 81, pp. 109–139, 1926.

[8] Howard Carmichael, *An Open Systems Approach to Quantum Optics*, Lecture Notes in Physics Monographs, vol. 18, Springer, 1993.

[9] Jacobs, K. & Steck, D.A., “A straightforward introduction to continuous quantum measurement,” *Contemp. Phys., vol.* **47**, pp. 279–303, 2006.

[10] G. Wentzel, “Eine Verallgemeinerung der Quantenbedingungen für die Zwecke der Wellenmechanik,” *Z. Phys.*, vol. 38, pp. 518–529, 1926.

[11] H. A. Kramers, “Wellenmechanik und halbzahlige Quantisierung,” *Z. Phys.*, vol. 39, pp. 828–840, 1926.

[12] L. Brillouin, “La mécanique ondulatoire de Schrödinger: une méthode générale de résolution par approximations successives,” C. R. Acad. Sci., vol. 183, pp. 24–26, 1926.

[13] H. Jeffreys, “On certain approximate solutions of linear differential equations of the second order,” Proceedings of the London Mathematical Society, vol. s2-23, no. 1, pp. 428–436, 1924.

[14] J. B. Keller, “Corrected Bohr–Sommerfeld quantum conditions for nonseparable systems,” *Ann. Phys.*, vol. 4, pp. 180–188, 1958.

[15] Y. Kurihara, “Symplectic structure for general relativity and Einstein–Brillouin–Keller quantization,” *Class. Quantum Grav.*, vol. 37, no. 23, p. 235003, 2020.

[16] M. C. Gutzwiller, *Chaos in Classical and Quantum Mechanics*. Berlin, Germany: Springer, 1990.

[17] E. Schrödinger, “Der stetige Übergang von der Mikro- zur Makromechanik,” *Naturwissenschaften*, vol. 14, pp. 664–666, 1926.

[18] C. M. Bender, *PT Symmetry in Quantum and Classical Physics*. London, U.K.: World Scientific Publishing Europe Ltd., 2018.

[19] W. Ritz, “Über die kombinatorische Natur der Spektrallinien,” *Ann. Phys.*, vol. 25, pp. 901–948, 1908.

[20] D. Hestenes, “Quantum mechanics of the electron particle-clock,” arXiv:1910.10478, 2019.

[21] D. Hestenes, "Gyromagnetics of the electron clock," *IEEE Access*, 2024.

[22] L. de Broglie, *Recherches sur la théorie des quanta*. Ph.D. dissertation, Univ. Paris, 1924.

[23] L. de Broglie, "La nouvelle dynamique des quanta," in *Electrons et photons, Rapports et discussions du cinquième Conseil de Physique tenu à Bruxelles du 24 au 29 octobre 1927*, B. Despaigne, ed. Paris, France: Gauthier-Villars, 1928, pp. 105–132.

[24] L. de Broglie, *La mécanique ondulatoire*. Paris, France: Gauthier-Villars, 1928.

[25] L. de Broglie, *Monsieur De Broglie*. INA Archives, 1967. Available: https://www.ina.fr/ina-eclaire-actu/video/afe04002106/monsieur-de-broglie

[26] D. Hestenes, "The zitterbewegung interpretation of quantum mechanics," *Found. Phys.*, vol. 20, pp. 1213–1232, 1990.

[27] W. Heisenberg, "Über quantentheoretische Umdeutung kinematischer und mechanischer Beziehungen," *Z. Phys.*, vol. 33, pp. 879–893, 1925.

[28] M. Born and P. Jordan, "Zur Quantenmechanik," *Z. Phys.*, vol. 34, pp. 858–888, 1925.

[29] D. Bohm, *Wholeness and the Implicate Order*. London, U.K.: Routledge, 1980.

[30] J. Barandes, "The stochastic-quantum correspondence," arXiv:2302.10778, 2023.

[31] J. Barandes, "New prospects for a causally local formulation of quantum theory," arXiv:2402.16935, 2024.

[32] C. Jaimungal, "The 21st century's greatest living scientist | Roger Penrose," *Theories of Everything*, YouTube, Oct. 2024. [Online Video]. Available: https://www.youtube.com/watch?v=sGm505TFMbU

[33] D. Bohm, "A suggested interpretation of the quantum theory in terms of 'hidden' variables I," *Phys. Rev.*, vol. 85, pp. 166–179, 1952.

[34] E. Nelson, "Talk," Dept. Math., Princeton Univ., 2005. Available: https://web.math.princeton.edu/~nelson/papers/talk.pdf

[35] M. Nagasawa, *Schrödinger Equations and Diffusion Theory: Reprint of the 1993 Edition*. Basel, Switzerland: Birkhäuser, 1993.

[36] S. Kocsis, B. Braverman, S. Ravets, M. J. Stevens, R. P. Mirin, K. Shalm, and A. M. Steinberg, "Observing the average trajectories of single photons in a two-slit interferometer," *Science*, vol. 332, p. 1170, 2011.

[37] S. Kocsis, B. Braverman, S. Ravets, M. J. Stevens, R. P. Mirin, K. Shalm, and A. M. Steinberg, "Observing the average trajectories of single photons in a two-slit
interferometer," National Institute of Standards and Technology (NIST), 2014. [Publicly available reprint]. Available: https://tsapps.nist.gov/publication/get_pdf.cfm?pub_id=907599

[38] L. V. Hedges and I. Olkin, *Statistical Methods in Meta-Analysis*. New York, USA: Academic Press, 1985.

List of Figures

1 Introduction

The article is organized in two parts. Chapters 1–5 establish the formal basis, beginning with the Schrödinger equation and its semiclassical approximations. Chapters 6–11 shift to a broader discussion, combining historical reconstruction, critique of the standard axioms, and the proposal of an extended temporal framework. Each part may be read independently, but together they trace a continuous line of argument.

Physics has always moved within a tension between two extremes: on the one hand, the technical view of experimental physics, which claims that only what is measurable exists. On the other hand, the theoretical view, sharpened by Einstein's remark: "It is the theory which decides what we can observe." Without a theoretical framework, data remain meaningless, and a theory that never leads to testable consequences is not physics. From this standpoint, we advocate a simple but far-reaching attitude: fundamental equations that are empirically confirmed must not be regarded merely as calculation rules. They must also be read and understood in their mathematical structure—as an expression of physical relations. Anyone who does not share this orientation will struggle with the ideas presented in this article.

The longstanding conceptual tensions in quantum theory are well known and persist despite the formal success of its predictions. The superposition principle challenges classical logic; the notion of wave function collapse introduces discontinuity without dynamics; entanglement defies spatial separability; and probabilities are treated as irreducible, resisting causal reconstruction. While mathematically precise, QM appears to lack a coherent ontological structure. From our perspective, this deficit arises from a fundamental misinterpretation of the role of time. Our goal is not to propose a new dynamical theory, but to develop a conceptual framework that allows QM to be reinterpreted without altering its core mathematical machinery. We aim to offer an architecture in which superposition, collapse, and probability gain structural meaning — not through metaphysical speculation, but via a realignment of formal elements already present in the theory.

The decisive point is the fundamental asymmetry between time and space. Unlike space, time is irreversible: once a moment has passed, it cannot be revisited. Such irreversibility is absent from the equations of motion, which describe only reversible processes. In QM, however, the unitary evolution of the wave function—as a causal, reversible development—stands in direct contradiction to the nature of measurement. The two concepts are incompatible and expose the overextension of a mechanical paradigm into a domain where real events are fundamental. By *event* we mean not merely a state change or the manifestation of a mechanical property, but an occurrence that constitutes a structural break or the sudden emergence of a random influence. And is not time, in its very essence, inseparably tied to such events?

Einstein's decisive achievement was to demonstrate that space and time cannot be separated, a result that emerged from the structure of electrodynamics and was mathematically expressed in the four-vector formalism. At first sight, this unification appears tied to velocity, since Lorentz transformations quantify the dependence of temporal intervals on relative motion. Yet this is merely a matter of parametrization, not of principle. The inseparability of space and time holds universally. A structural comparison makes this even clearer: the phase of an elementary wave combines energy, momentum, and space–time coordinates in precisely the same form as the relativistic four-vector. From our perspective, the wave function must therefore be understood as an expression of this deeper unification. What physical clocks measure—Einsteinian time t—is only an ordering parameter within this broader spatio-temporal structure, whose coherence is encoded in the wave function itself. From this perspective, the essential difference between relativity and QM lies not in whether space and time are unified, but in how this unity manifests: in relativity through mechanics, and in QM through probabilistic structure. It is therefore natural to review QM from a temporal point of view, explicitly including non-local time structures.

A precursor to this view can be found in de Broglie's theory—where quantum behavior emerges from the coherence between one internal vibration and its relativistic transform—and here and throughout, we understand "events" as manifestations of complex, fundamentally probabilistic processes. This resonance-based perspective implicitly bridges SR and QM and motivates our interpretation of time as *holoperspectival* (i.e., reflecting both local and global scales). We refer to this ordering principle as *well-tempered time* (analogous to the deliberate slight mistuning of intervals in *well temperament* tuning in music). In such a framework, time acquires an additional degree of freedom—akin to proper time in relativistic formulations. The wave function, traditionally treated as a probability amplitude, is instead seen as an indivisible complex flow of dynamic time, carrying both phase and amplitude.

This motivates a reformulation of the wave function as a temporally structured object, shaped by resonance and governed by internal phase dynamics. The Schrödinger equation then describes not the evolution of probabilities, but the viability of a time-structured flow.

Beyond its conceptual coherence, this view opens the door to empirical investigation. We outline how phase-resolved measurements and temporal coherence patterns in quantum systems could serve as experimental indicators of structured time. These proposals suggest that structured time is not a metaphysical supplement, but an empirically accessible layer of physical reality.

To develop this approach, we proceed in five steps: First, we reinterpret complex numbers as expressions of time structure. Second, we recast the wave function as a flow of such structured values. Third, we reassess the axioms of quantum theory from this temporal perspective. Fourth, we identify points of contact with experiment.

This framework is offered as a foundation for rethinking quantum theory — not as an overhaul, but as a reorientation grounded in the theory's own formal capacities.

Some elementary aspects of QM are discussed in greater detail than usual, even at the level of basic topics. This is necessary because our unconventional perspective of QM as temporal flow requires a systematic review of the evidence."

2 Evaluated Time and Time Numbers

As stated above, Einstein's achievement was to establish the inseparability of space and time, a principle that we extend to QM by noting the structural analogy between the relativistic four-vector and the phase of a wave function. This perspective implies that what clocks measure—Einsteinian time t—is only an ordering parameter, while the coherence of space-time is encoded directly in the wave function. Building on this, we now search for a consistent extension: the wave function already contains an implicit generalized time structure. Our starting point, therefore, is to examine the wave function itself in greater detail. The most straightforward way to begin is by analyzing its values—namely, the complex numbers on which it is built.

This section condenses [1], highlighting the key passages most relevant here. However, we present the motivation in an unusually detailed manner, since even today complex numbers largely lack an acknowledged meaning.

From school mathematics, we are familiar with natural numbers for counting, integers for ordering, rational numbers for expressing proportions, and real numbers for measuring continuous magnitudes. These number systems reflect successive levels of abstraction and functional scope.

Complex numbers are not just formal devices. In polar form, the modulus encodes weight, while the phase marks temporal position and polarity. A nonzero phase φ indicates latent dynamics, while $\varphi=0$ corresponds to observable events. Superposition thus expresses the coexistence of latent and manifest components, without mysticism.

Motivation

We seek to understand complex numbers as a representation of qualities with opposing manifestations—for example, account balances in the form of credit versus debt. As a first simple step, clarifying multiplication on the real line may help.

Let a, b > 0. Then:

- $(-a)b = (-1)ab$: scaling by a combined with a π rotation (i.e., a reflection across zero), corresponding to a reversal of quality.
- $(-a)(-b)= (-1)(-1)ab$: two π rotations yield a full 2π rotation, effectively reversing
  the quality twice—and thereby restoring the original orientation.

This suggests the rule: moduli multiply, phases add—a continuous modulation of quality along the phase. Phases $\varphi \neq 0°, \pi$ may indicate that a process is ongoing—potentially leading to a payout (first and fourth quadrants) or a write-off (second and third), depending on the temporal direction. Superposition thus becomes a technical term for combining such qualities—particularly when the phase $\varphi \neq 0°, \pi$ reflects hidden, preparatory, or non-monetary processes such as negotiations or internal restructuring. We emphasize that this is not merely a matter of incomplete knowledge. Rather, it reflects an intrinsic feature of temporally extended processes: they cannot be fully captured by static assignments and require a representation that accommodates internal differentiation—idealized here as superposition. From this, the concept of an expectation value arises naturally: a concrete realization corresponds to a value triggered—but not determined—by probability.

Thus: This enables a continuous comparison within a single quality category. More general cases will be treated later.

In contrast to the Newton–Einsteinian view, we describe observations as pairs (b,φ), both depending on x and t. Here b denotes the weight or significance of the event (e.g., probability, intensity), while φ marks its temporal phase—the position of the quality in the unfolding of time. Importantly, b need not be normalized. We call such pairs **time numbers**.

After a broad and illustrative motivation, this section now turns to the precise definitions.

Definition 2.1

1. An ordered pair (b,φ) with real numbers φ and $b > 0$ is called an *Evaluated Time number* (*Zeitzahl*).
2. A function ψ, assigning to each position x and Einstein time t a *time number* $(b(x,t),\varphi(x,t))$, is called an *Evaluated Time Measure* (*Zeitmaß*).

To be able to talk about numbers at all, we need a meaningful way to compare time numbers. In such a comparison, both components—weight b and phase φ—must enter jointly; otherwise, we would merely have two isolated real numbers. Comparisons presuppose categories $K_1, K_2, \ldots, K_N$, within which such time numbers are comparable, while categories themselves remain distinct. When elements from different categories appear together, we can formally write tuples such as (z_i, z_j). For

N categories this yields an N-tuple, and for $N=\infty$ a countably infinite tuple, i.e. a sequence. Extending this idea suggests a further step: one may form combinations like $z_i \otimes z_j$, reminiscent of tensor products, which in ordinary mathematics encode independent degrees of freedom. The point is that considerations of comparability and composition lead naturally toward the arithmetic structure familiar from Hilbert spaces.

For numbers within the same category, we stipulate: let $z_1=(b_1,\varphi_1)$ and $z_2=(b_2,\varphi_2)$. Their multiplication is defined by multiplying the moduli and adding (or subtracting) the phases, consistent with the representation $z = be^{i\varphi}$, where moduli multiply and phases combine in the exponent. This structure mirrors harmonic relations familiar from mechanics (gear ratios) and from music theory (the overtone series as integer multiples of the fundamental).

Addition, by contrast, is immediate: thinking in terms of vector addition, the result lies closer to the time number with the larger modulus b.

Thus, while tensor products of Hilbert spaces may be mathematically indispensable, our discussion shows that they merely represent simple temporal structures. What appears as abstract algebra is, at its core, the straightforward composition of time qualities. Once this is recognized, the question of dimensionality also gains clarity: Hilbert spaces appear in finite or infinite dimensions because each eigenvalue corresponds to a distinct quality (measure). To capture all such components simultaneously, they must be separated in the representation, since different measures are incommensurable, and this separation is precisely what introduces multiple dimensions. In this sense, the passage to infinite-dimensional Hilbert spaces—physically always understood in the separable case with countably infinite bases—appears naturally as the step that includes all possible spectral qualities within a unified temporal framework, for example.

Definition 2.2:

Given two time numbers $z_1 = (r_1,\Phi_1)$ and $z_2 = (r_2,\Phi_2)$, then

1. their sum is

   $z = z_1 + z_2 = z1+z2=(r_1\cos \Phi_1+ r_2\cos \Phi_2, r_1\sin \Phi_1+ r_2\sin \Phi_2)$ and

2. their product is

   $z = z_1 \cdot z_2 = (r_1 \cdot r_2,\Phi_1 + \Phi_2)$.

Definition 2.3 (Generalized Born's Rule)

Time-numbers z_1 and z_2 are evaluated by the *(generalized) Born rule*:

$$\langle z_1|z_2\rangle = z_2^* \cdot z_1 = (r_1 \cdot r_2,\Phi_1 - \Phi_2). \qquad (2.1)$$

Comparison thus amounts to multiplying the complex conjugate z_2^* with z_1, i.e., by dividing z_1 by z_2. This procedure can also be applied pointwise to evaluated time measures, where available. For a bilinear extension of Born's rule related to the present ansatz (2.1), see [2].

In purely mechanical theories, evolution reduces to deterministic state transitions—mere reparameterizations without genuine "becoming"—so that mechanics is, at its core, timeless. SR already modifies this by binding proper time (Einsteinian Time) inseparably to each world-line: one cannot speak of duration without reference to motion itself. Building on this inseparability of time and process, we propose to extend the same principle to systems whose internal complexity forces part of their dynamics into a stochastic description. Sudden, non-mechanical transitions—process events—break through deterministic flow and reveal an intrinsic process-time: Evental Time.

Definition 2.4

- **Einsteinian Time** refers to the continuous, mechanical time parameter in classical and relativistic physics. It is measured by local clocks along worldlines and treats all processes as reversible trajectories in state space—thereby excluding any notion of genuine events or irreversible emergence.
- **Event:** A discrete occurrence or manifestation of a fundamentally or apparently probabilistic process.
- **Eventual Time:** We assume that even apparently random events express an underlying (non-local) order, whose manifestations we regard as expressions of Eventual Time.

In this article, the term "Eventual Time" is not intended to introduce a transcendent or mystified notion of time. Rather, it has a dual reading. On the one hand, in the narrow, technical sense (developed further in Part II), it amounts to a theory of temporally structured probabilistic measures or processes—directly operational and statistically testable. On the other hand, the term signals openness: a recognition that Einsteinian time is too narrow, and that excessive precision may itself become a constraint, too rigid a corset to capture reality. The two readings are not contradictory but complementary: the first expresses what is concretely achieved in this work, while the second preserves conceptual space for a more adequate philosophical interpretation elsewhere. Therefore, from our perspective, Einsteinian Time—and the mechanical worldview it supports, including classical and relativistic mechanics—appears as continuous and deterministic, yet fundamentally blind to events.

This distinction can be illustrated by a simple analogy: like a film, where Einsteinian Time merely counts the scene numbers, while Evental Time follows the dramatic twists of the story.

Since locality is a minimal requirement for any mechanical theory, *Evental time* cannot be local. It instead emerges from the process as a whole and its coherence with global structures—first revealed in quantum entanglement. Non-locality simultaneously suggests a holographic principle: the self-similarity of structures across different scales allows us to regard the process as its own optimal clock and to endow it with a coarse-grained, globally encompassing temporal structure.

From this vantage, SR (and its implied Block Universe) remains event-less, while QM—through its evaluated time measures (e.g. the wave function ψ)—maps out a richly event-driven evolution. The simplest mathematical framework for capturing Evental Time is precisely provided by evaluated time measure.

High-significance evidence for the existence of Evental Time has been presented in [3,4]. In the following sections, we analyze key aspects of QM through the lens of time, focusing on how evaluated time measures reshape our understanding of quantum phenomena. In Chapter 10, we then propose concrete experimental tests for *Eventual Time* within quantum-mechanical frameworks.

3 Review of Schrödingers Equation

The central equation of nonrelativistic QM is the time-dependent Schrödinger equation (SE for short) [5]:

$$i\hbar\frac{\partial\psi}{\partial t} = (-\frac{\hbar^2}{2m}\nabla^2 + V)\psi. \qquad (3.1)$$

The expression

$$\hat{H} = -\frac{\hbar^2}{2m}\nabla^2 + V \qquad (3.2)$$

on the right-hand side, which acts on ψ, is called the Hamiltonian operator. Using this notation, the Schrödinger equation takes the more compact form:

$$i\hbar\frac{\partial\psi}{\partial t} = \hat{H}\,\psi \qquad (3.3)$$

and, for the stationary case [6],

$$\widehat{H}\,\psi = E\psi. \tag{3.4}$$

Rearranging (3.3) to isolate the time derivative yields

$$\frac{\partial\psi}{\partial t} = \frac{i}{\hbar}\widetilde{\widehat{H}}\psi,\ \widetilde{\widehat{H}} = \frac{\hbar^2}{2m}\nabla^2 - V. \tag{3.5}$$

Equation (3.5) shows a first-order time derivative on the left-hand side, whereas the right-hand side exhibits a formal similarity to a diffusion generator and—via the continuity law—to a flow equation. Both sides therefore describe motion, giving the Schrödinger equation a consistent dynamical interpretation.

Further interpretation will be supplied by Section 9.2.

Rather than being a mere calculational device, the complex notation is essential. It compactly encodes that the Schrödinger equation is not a single real equation but a system two intrinsically coupled real equations. This becomes explicit when rewriting the complex structure in real matrix form (the arrow '↦' denotes the representation of complex numbers in real coordinates):

$$u + iv \mapsto \begin{pmatrix} u & -v \\ v & u \end{pmatrix}.$$

Multiplication by i is represented in real coordinates by the matrix

$$J = \begin{pmatrix} 0 & -1 \\ 1 & 0 \end{pmatrix} \tag{3.6}$$

By writing $\psi = u + iv$, the SE takes the following representation:

$$\frac{\partial}{\partial t}\begin{pmatrix} u \\ v \end{pmatrix} = \frac{1}{\hbar}\begin{pmatrix} 0 & -1 \\ 1 & 0 \end{pmatrix}\widetilde{\widehat{H}}\begin{pmatrix} u \\ v \end{pmatrix}. \tag{3.7}$$

In (3.7), the real and imaginary parts of ψ—namely u and v—are necessarily coupled, because the purely off-diagonal form of J enforces their mutual dependence. Hence, the Schrödinger equation cannot be split into two independent real equations.

The apparent oscillations stem from the complex phase; interpreting the Schrödinger equation as a classical oscillatory (hyperbolic) wave equation is therefore a misrepresentation.

4 Elementary Examples

We continue our foundational considerations by examining simple examples. At the same time, we are aware that in QM situations often appear no less ambiguous than in the macroscopic world. Even mesoscopic flows, where energy transport tends to favor wave-like behavior, frequently display such properties. For instance, when a water wave impinges on an obstacle, the flow may split into a superposition of two counter-propagating waves, so that the phenomenon takes on a wave-like appearance.

To avoid becoming entangled in misleading debates, we acknowledge this issue but deliberately restrict ourselves in what follows to simple and clear examples. Subsequently (Sections 5.1 and 5.2), we substantiate the inherent circularity by further considerations, in particular through the EBK approximation. There, the classical quantization condition emerges directly, which places our perspective into an established context. Alongside the formal analysis of the Schrödinger equation (Eq. 3.5) and the discussion of norm preservation (Section 5.1), this provides an additional, independent argument.

4.1 Free Particle

Since our discussion centers on temporal flow, we begin with the time-dependent Schrödinger equation. For a free particle moving along the x-axis,

$$i\hbar\frac{\partial\psi}{\partial t} = -\frac{\hbar^2}{2m}\frac{\partial^2}{\partial x}\psi.$$

Its solutions are plane waves

$$\psi(x,t) = Ae^{i(kx-\omega t)} + \ Be^{-i(kx+\omega t)}, \omega = \frac{\hbar k^2}{2m}.$$

For visualization, we restrict ourselves to the forward-traveling component

$$\psi(x,t) = Ae^{i(kx-\omega t)}.$$

In this simple example, considering motion along the x-axis with constant velocity $v = \hbar k/\mathrm{m}$, we set $x(t) \ = vt$. Substituting this into the wave function gives a helical curve in three dimensions (see Figure 1),

$$\gamma(\mathrm{t}) = (\boldsymbol{x}(\mathrm{t}), \Re\psi(x(t),t), \Im\psi(x(t),t)),$$

where $\Re\psi$ and $\Im\psi$ denote the real and imaginary parts of the wave function, respectively. This representation makes the temporal circulation explicit. Unlike conventional depictions that separate modulus, real, and imaginary parts, our perspective emphasizes the flow of two-component evaluated time measures introduced in Chapter 2. In contrast to phasors in engineering, the components of ψ are not redundant: its complex structure is irreducible and essential in QM. The familiar decomposition into periodic cos and sin functions is only a visualization; it conceals the genuine phase rotation in the complex plane.

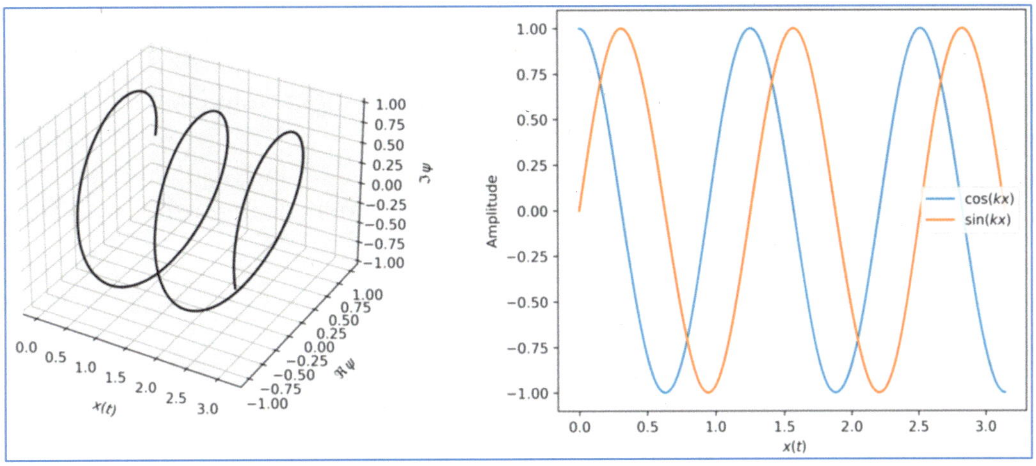

Figure 1: Free particle–curve γ (left) and its decomposition into cosine and sine components (right).

4.2 A More General Example with Nonvanishing Potential V

In contrast to the free particle case, where the wave function exhibits a single-frequency circulation in the complex plane, the situation becomes richer once a potential is present. As a representative example, consider the harmonic oscillator. At position x, the wave function can be expanded in eigenstates

$$\psi(x,t) = \sum_n C_n \phi_n(x) e^{-\frac{iE_n t}{\hbar}}, \quad E_n = \hbar\omega\left(n + \frac{1}{2}\right)$$

Time evolution then involves several frequencies: the curve traced in the complex plane is no longer a perfect circle but reflects the interference of components. For coherent states the picture simplifies [7]: the probability maximum coincides with the centroid, and the packet follows the classical orbit without spreading.

For visualization, we again employ the representation introduced in Section 4.1, reducing the dynamics to a single spatial path given by the centroid $x(t)$ and plotting the real and imaginary parts of ψ along the other two axes. This yields the closed curve

$$\gamma(t) = (x(t), \Re\psi(x(t),t), \Im\psi(x(t),t))$$

which makes the temporal circulation explicit. As discussed in Section 4.1, a separated depiction of modulus, real, and imaginary parts would miss the essential point. Our perspective instead emphasizes the flow of two-component evaluated time measures (b,φ) introduced in Chapter 2; decomposing them would contradict their very definition.

Figure 2 illustrates a closed curve, with its four quarters distinguished by different color intensities.

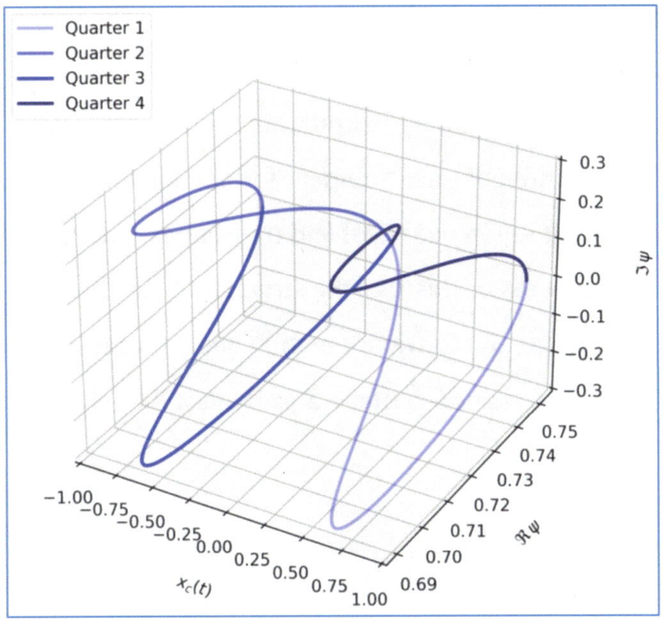

Figure 2: Harmonic oscillator—curve γ.

5 The Wave Function as a Circular Flow

Can the Schrödinger equation, beyond the simple cases considered in the last section, still be interpreted as a flow akin to diffusion? One might suspect instead that the wave function, when taken in full generality, could blur into a diffuse block, much like a Fourier superposition without intrinsic structure. This classical viewpoint treats the wave function as nothing but an aggregate of components. Yet both modern continuous-measurement theory and well-established approximation schemes such as WKBJ and EBK demonstrate otherwise: the circular term is essential. It is not washed out by superposition, but persists as a defining element of the wave function's structure, and it plays a central role in practical applications.

Modern approaches in quantum optics and quantum information address this issue by treating continuous measurement as a stochastic process, formulated through stochastic Schrödinger equations and so-called quantum trajectories ([8],

[9]). In this framework, the reduced state is typically represented as a Gaussian-like packet, reflecting spatial localization, while the underlying unitary phase evolution remains intact. This dual picture — localization and circulation — resonates with the structure we propose here. This continuous-measurement formalism will not be pursued here in detail. Instead, we turn to more classical and transparent examples, which have long been known and allow us to illustrate the structure more directly. Sections 5.1 and 5.2 revisit the circulation of the wave function and its representation in WKBJ- and EBK-approximations, while Section 5.3 presents a first elementary non-Hermitian example.

5.1 Circulation of the Wave Function Due to Norm Conservation

From the time-dependent Schrödinger equation

$$i\hbar \frac{\partial \psi}{\partial t} = \hat{H}\psi$$

it follows immediately

$$\frac{d}{dt}\langle\psi|\psi\rangle = 0,$$

Equivalently, the squared norm remains constant,

$$\frac{d}{dt}\|\psi(\mathrm{t})\|^2 = 0,$$

i.e. the norm ψ, being just the square root, is conserved. The time-evolution opera

tor (often called the propagator) is therefore given by

$$U(t,t_0) = e^{-\frac{i}{\hbar}H(t-t_0)},$$

which is unitary. For a small time step dt, the state updates as

$$\psi(t+\mathrm{d}t) = U(t+\mathrm{d}t,t)\psi(\mathrm{t})$$

which geometrically corresponds to a rotation in Hilbert space (see Fig. 3).

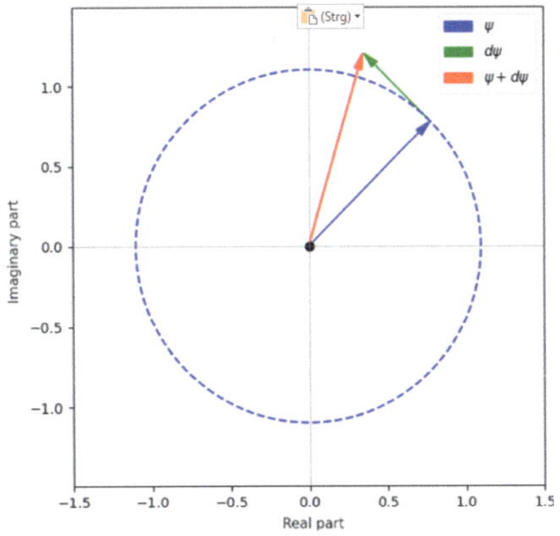

Figure 3: ψ(t+dt) ≈ ψ(t)+dψ

In the linear approximation $\psi(t+\mathrm{d}t) \approx \psi(\mathrm{t}) + \mathrm{d}\psi$, the updated state does not yet exactly preserve the norm; only the full exponential

$$U = e^{-\frac{i}{\hbar}Hdt}$$

restores unitarity. This is why the approximated vector in Fig. 3 lies slightly outside the unit circle

5.2 Circular Flow Shown by WKBJ- and EBK-Approximation

In this section we deepen the idea that the wave function represents a spatially extended complex function with a temporally circular component. To this end, we employ two well-established approximation methods. The aim is to obtain information about the phase of the wave function also in the general case—analogous to the situation of a free particle, where this is easy to see. This will enable us, in Section 6.5, to justify our view of the wave function as an evaluated measure of time, extending beyond an adequate mathematical interpretation to a sufficient physical grounding.

5.2.1 WKBJ-Approximation

The technique now known as the WKBJ approximation was developed independently in 1926 by G. Wentzel [10], H.A. Kramers [11], L. Brillouin [12], and H. Jeffreys [13].

The essence of their approach is to represent the wave function in semiclassical form, with an amplitude and a rapidly varying phase governed by the Hamilton–Jacobi equation. Concretely, when one treats the full time-dependent Schrödinger equation, one assumes the ansatz

$$\psi(x,t) \sim A(x,t)\, e^{\frac{i}{\hbar}S(x,t)},$$

with $S(x,t)$ satisfying the Hamilton–Jacobi equation with time,

$$\frac{\partial S}{\partial t} + H(x, \nabla S) = 0.$$

In this case, S explicitly depends on t. WKBJ is a local semiclassical method. The phase function S plays the role of the classical action, with

$$\nabla S = \boldsymbol{p},\, E = -\frac{\partial S}{\partial t},$$

identifying momentum $\boldsymbol{p}$ and energy E via Hamilton–Jacobi theory.

Thus, a first-order Taylor expansion around a reference point yields locally

$$S(x,t) \approx \mathbf{p}\cdot\mathbf{x} - Et,$$

so that the wave function reduces to the form of a plane (elementary) wave.

Importantly, this semiclassical approach was developed independently of Madelung's hydrodynamic formulation of QM, although both employ the same amplitude–phase decomposition. WKBJ, however, shows constructively why the phase must act as the essential dynamical variable, rather than relying on interpretation alone.

5.2.2 EBK-Approximation

For a Liouville-integrable system with f degrees of freedom, the semiclassical circularity of the wave function is expressed by the EBK (Einstein–Brillouin–Keller) rules. On each independent closed cycle γ_j of the classical invariant torus ($j = 1,\ldots,f$) one has (see Keller [14]; Kurihara [15])

$$\oint_{\gamma_i} \boldsymbol{p} \cdot d\boldsymbol{q} = 2\pi\hbar\left(n_j + \frac{\mu_j}{4}\right).$$

Here n_j are integer quantum numbers, and μ_j are Maslov indices accounting for phase jumps at turning points or caustics. In the one-dimensional case ($f = 1$) with two turning points, $\mu = 2$, and the rule reduces to the Bohr–Sommerfeld quantization with Maslov correction:

$$\oint p\, dq = 2\pi\hbar\left(n + \frac{1}{2}\right).$$

If the dynamics can be separated into canonical pairs the EBK condition takes the reduced form

$$\oint_{\gamma_i} p_j\, dq_j = 2\pi\hbar\left(n_j + \frac{\mu_j}{4}\right), j=1,\ldots,f.$$

The EBK approximation emphasizes the inherently circular character of quantum wave functions. It reinforces the view that quantum dynamics is fundamentally circulatory rather than merely wave-like. Its formulation in terms of symplectic and topological indices has been shown to extend even to relativistic systems [7], underlining the universal character of the underlying structure.

For systems that are not Liouville-integrable, the EBK rules no longer apply. In this regime, semiclassical quantization is instead governed by the Gutzwiller trace formula (see [16]), which expresses the density of states as a sum over contributions from unstable periodic orbits. This formulation makes explicit that local predictability is fractured in chaotic systems. Yet, the wave function remains globally organized through the collective contributions of its periodic orbits. Taken together, the EBK and Gutzwiller frameworks demonstrate that even when local predictability breaks down, QM retains a structural coherence—through toroidal cycles in the integrable case, or through the web of periodic orbits in the chaotic case. What first appeared with Bohr thus resurfaces across semiclassical frameworks, pointing to a deeper structural principle.

The EBK rules themselves had their precursors in the WKBJ approximation, originally stimulated by Schrödinger's note [17].

5.3 Flows with Drift

For interpreting the wave function as a circular flow in 5.1, unitarity is not a necessary condition. It is merely sufficient, since it guarantees that the process reduces to a pure infinitesimal rotation. Yet beyond this ideal case, a dominant drift is equally conceivable and will be discussed in this section. We begin with a simple example from [18], where the Hamiltonian is defined as

$$\hat{H} = \mathrm{a} + \mathrm{ib},$$

with a and b real constants, b≠0, and i the imaginary unit. Clearly, the Hamiltonian $\hat{H}$ is not Hermitian, since

$$\hat{H}\text{-}\hat{H}^* = 2\mathrm{ib} \neq 0.$$

We now write the Schrödinger equation in units where $\hbar = 1$:

$$\frac{\partial\psi}{\partial t} = -i(\mathrm{a+ib})\psi = (\mathrm{b\text{-}ia})\psi.$$

Its solution is given by

$$\psi(t) = \psi(0)e^{(b-ia)t} = \psi(0)e^{bt}e^{-iat}.$$

Thus, the wave function separates into a product of a real exponential function e^{bt} and a circulation e^{-iat} on the unit circle in the complex plane, scaled by the initial value $\psi(0)$, the wave function at t = 0. The dominant drift term e^{bt} diverges for b>0 and decays to zero for b<0.

Figure 4 illustrates the case a = $-(2\pi)^{-1}$ and b = 1; it describes a spiral winding outward from the origin.

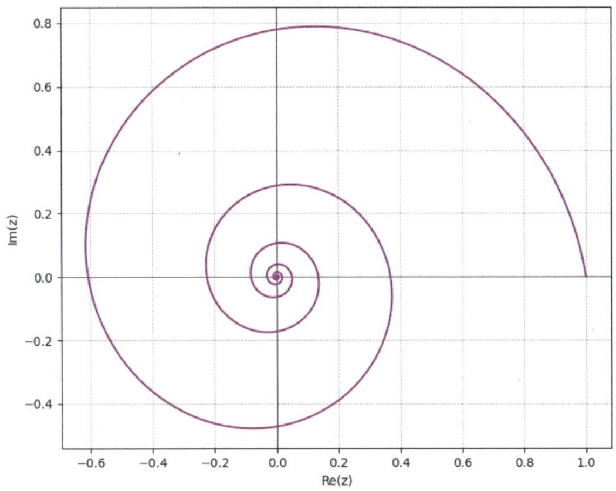

Figure 4: Wave function for $\widehat{H} = -(2\pi)^{-1} + i$

Bender ([18]) discusses much more general examples relevant to *important experiments, for instance in quantum optics*. Instead of Hermiticity, PT symmetry is assumed; this is known as PT-symmetric QM.

Already a slight generalization of the simple model above, namely a system with a 2×2 Hamiltonian

$$\widehat{H} = \begin{pmatrix} a + ib & g \\ g & a - ib \end{pmatrix}$$

exhibits remarkable properties: $\widehat{H}$ is PT-symmetric and can have real eigenvalues even though it is clearly not Hermitian. Here, $\widehat{H}_{11}$ and $\widehat{H}_{22}$ represent subsystems with energy gain and loss at rate b, respectively, while the coupling g mediates the exchange between them.

Even if fundamental axioms of QM, such as Hermiticity, are relaxed, the notion of circular flow persists, merely extended by a drift component. This highlights the robustness of the circulatory interpretation beyond the standard framework.

6 Time Structure in Quantum Theory: From Ritz to Hestenes

In the early years of quantum theory, Walter Ritz and Louis de Broglie placed aspects of temporal structures at the center of attention. With the Fifth Solvay Conference in 1927, focus shifted toward energy and probability alone. We briefly outline this development below.

6.1 Ritz's Combination Principle as Time Geometry

Walter Ritz (see [19]) noted in 1908 that spectral lines satisfy

$\nu_{mn} = T_m - T_n,$

where each T_n denotes a frequency—the expression of an intrinsic temporal rhythm, rather than merely the energy associated with it. Such frequencies reflect the discrete temporal rhythms of atomic processes. Their pairwise differences capture, from our perspective, a network of temporal structures: a set of 'inner clocks' linked by frequency coherence. Although Ritz did not pursue this line of thought consistently, the potential is nevertheless evident, as Hestenes later demonstrated [20,21].

6.2 De Broglie's Coherence Condition, Not a Matter-Wave Duality

Contrary to the common wave–particle duality myth, de Broglie never claimed that matter itself *is* a wave. Inspired equally by Einstein's light–quantum hypothesis and the invariance principles of special relativity, he imposed a coherence requirement between a particle's proper–time oscillation

$$f = \frac{E}{h} \qquad (6.2.1)$$

and a phase wave in the laboratory frame via the appropriate Lorentz transformation [22]. From this single insight he obtained the de Broglie relation

$$\lambda = \frac{h}{p} \tag{6.2.2}$$

and, in consequence, a reinterpretation of the Bohr–Sommerfeld quantization rule

$$\oint p\, dq = nh. \tag{6.2.3}$$

which now appeared as a natural phase–closure condition.

In de Broglie's reasoning, two different frequencies appear in the laboratory frame. The time-dilated inner vibration ν_0 of the particle runs with $\nu_0\sqrt{1-\beta^2}$, while the Planck–Einstein relation applied to the relativistic energy gives $\nu_0/\sqrt{1-\beta^2}$. One decreases with velocity, the other increases. To prevent this divergence, de Broglie imposed a coherence condition: both aspects must agree.

He also derived a speed $V = c^2/v$ for the wave assigned to the particle, where v is the particle's velocity. As this speed exceeds the speed of light, de Broglie explicitly described the accompanying wave as fictitious [22]: "Cette onde de vitesse plus grande que c ne peut correspondre à un transport d'énergie; nous la considérerons seulement comme une onde fictive associée au mouvement du mobile." Thus, he characterized the wave as *fictitious*, while at the same time obtaining its wavelength $\lambda = h/p$. This result follows directly from the most elementary quantum input—the Planck–Einstein relation—together with special relativity. The wave nature of matter, and thus the starting point of QM, can therefore be seen as a direct consequence of Einstein's energy relation once consistency with relativistic time transformation is required.

Later misunderstandings partly stem from a mistranslation of de Broglie's *ondes de matière*, which more accurately means "waves associated with matter" rather than implying that matter itself propagates as a wave.

De Broglie worked in isolation and with remarkable courage. After facing harsh criticism at the 5th Solvay Conference in 1927 [123], he unexpectedly aligned with the Copenhagen school [24]—largely because the by-then dominant wave function picture had gained wide acceptance, while alternatives could scarcely prevail [9]. As de Broglie later emphasized [25], his derivation hinges on the full Lorentz symmetry—far more nuanced than most textbook accounts suggest.

Summary 6.2.1

De Broglie derived his famous relation (6.2.2) solely from special relativity, the basic energy formula for quanta and the assumption that the electron exhibits an internal oscillation—an assumption rather weak in light of the empirical reality of atomic spectra.

6.3 Hestenes' Clifford-Algebra Clock

David Hestenes [26] gives a concrete mathematical realization of the cyclic temporal quality posited here: the complex phase of the Dirac spinor is a real physical rotor in spacetime algebra whose internal phase rotates at the de Broglie angular frequency $\omega_B = mc^2/\hbar$, while the rapid circulatory motion (*zitterbewegung*) at $\omega_{zbw} = 2mc^2/\hbar$ produces spin and the magnetic moment. The "phase harmony" or coherence condition of de Broglie—synchronizing the internal clock with translational motion—emerges naturally from this unified rotor structure, not as an independent axiom. Spin is thus not merely encoded temporally; the internal clock and spin are two aspects of one underlying geometric motion.

In later work Hestenes ([20,21]) shows that electron mass is fully reduced to the clock frequency of its internal zitterbewegung—defining an "electron time" from that motion—and his exposition of the Dirac theory, closer to Dirac's original intention, places de Broglie's electron clock at the center as the physical mechanism unifying spin, mass, and phase coherence.

Sections 6.1–6.3 traced a deliberate progression from minimal formal observations to increasingly general structure: from Ritz's frequency geometry, through de Broglie's relativistic phase coherence, to Hestenes' internal clock. Keeping Schrödinger dynamics in view, the next step was only outlined: a minimal differential construction that extracts a temporal measure from phase via WKBJ/EBK elementary waves. A brief historical interlude will now follow, presenting Heisenberg's reinterpretation and the spectral foundation. Section 9.1 will then formulate our precise notion of time in QM as dynamical proper time.

6.4 Historical Context of Quantum Mechanics

Heisenberg [27], proceeding from Newton's law of force and inspired by the Fourier expansion of periodic motion, introduced the ansatz for the trajectory $x(t)$:

$$x(t) = \sum_{\alpha=\infty}^{+\infty} a_\alpha(n) e^{i\alpha\omega_n t},$$

where ω_n denotes the fundamental frequency associated with the Bohr–Sommerfeld quantum number.

The construction relied on five established elements:

1. **Newtons law of motion**: $\ddot{x} + f(x) = 0$
2. **Bohr–Sommerfeld quantization**: the integer n indexing the fundamental frequency ω_n.
3. **de Broglie's wavelength assignment:** $\lambda = h/p$, obtained by combining the Planck–Einstein relation with special relativity under the assumption of an internal vibration of the electron.
4. **Planck–Einstein relation:** $E = h\nu$.
5. **Ritz's combination principle of spectral frequencies**: $\omega_{mn} = \omega_{ml} + \omega_{ln}$.

From this perspective, the early formulation of quantum mechanics may be interpreted essentially as a theory of atomic spectra. Its scope extended no further than the observed regular occurrence of interference patterns and spectral lines.

The summation formula introduced by Heisenberg did not constitute a new postulate. Its significance lay in the systematic integration of these elements, which revealed the underlying algebraic structure subsequently recognized as matrix mechanics.

The involvement of matrices is already implicit in the additivity of frequencies, as formulated by Walter Ritz, since the right index of one term and the left index of the other must coincide. This index-matching corresponds precisely to the rule of matrix multiplication: the sum represents the matrix element (m,n) obtained through an intermediate state l.

Remark 6.4.1

Heisenberg's summation formula was derived exclusively from Newton's law of force, Ritz's combination principle of frequencies, the Bohr–Sommerfeld quantization condition, the Planck–Einstein relation between energy and frequency, and de Broglie's assignment of a wavelength to matter. Both the matrix formulation (Heisenberg–Born–Jordan) and the equivalent wave-function formulation (Schrödinger) thus rest on explicit assumptions which, in hindsight, may be regarded as transparent.

Subsequently, Schrödinger introduced his equation and demonstrated its equivalence to the Heisenberg formalism. The axiomatization that followed did not constitute a mere refinement. Rather, it amounted to a far-reaching generalization which, from the present perspective, had the effect of obscuring the spectral origin of the theory. This elevation of elementary quantum mechanics requires critical examination—a discussion to which we return in a subsequent section.

7 Hidden Variable and the Incompleteness of Quantum Mechanics – The Phase of the Wave Function

In the previous chapter, we highlighted a decisive shift in focus—from foundational considerations involving temporal structures to a purely spectral and computational treatment of energy levels. This misdirection began more than a century ago and has persisted to this day, despite paradoxes, unresolved contradictions, and the absence of a coherent ontology. In [3], based on the analysis of random data, we argued that the prevailing notion of time (Einsteinian Time) is insufficient to account for the emergence and structure of events.

Within that work, we introduced the idea of evaluated time measures, denoting the effects of complex or even stochastic processes as manifestations of a temporal unity. This perspective resonates with David Bohm's notion of the *Implicate Order* [29]. The central experimental tool is time selection (Zeitwahl), where measurements are performed only under specific temporal relations (see [3]; Sec. 11.3 in [4]; and Empirical Note 9.2.1). This principle applies at the level of single trials involving individual particles.

The present chapter develops the physical consequences of this view, focusing on the revision of quantum theory itself. From our perspective, the attempt to prove that hidden variables cannot exist was misguided—because Born's rule constructs probabilities in which the phase of the wave function is effectively lost. The conclusion is unavoidable: the phase of the wave function remains hidden—interpretable either as a hidden variable in its own right, or as the projection of a deeper, yet unknown one. Chapter 10 will support this view by presenting empirical evidence.

Moreover, the widespread rejection of hidden-variable models rests on Bell's analysis, which tacitly assumes a Markov-type structure of probabilities.

As Jacob Barandes has recently emphasized [30,31], such an assumption is not compulsory: quantum systems can be consistently modeled as indivisible, non-Markovian stochastic processes. This dissolves the supposed no-go for hidden variables and reopens the question on a more rigorous footing.

It is essential for interference and superposition, yet inaccessible to conventional physics. This very inaccessibility exposes the incompleteness of quantum theory—an incompleteness long masked by the dogma that randomness is intrinsic to QM.

If such far-fetched constructions (like Schrödinger's cat) are required to explain the Copenhagen interpretation, it seems far more coherent to instead assign an ontology to phases. One could then say: phases can be attributed to objects, allowing them to enter into superposition—not as objects themselves, but through their phases. Superposition thus expresses that two opposing events can potentially become real; it does not imply that objects exist in contradictory states. An event is characterized by its encoding in the system's phase—a non-mechanical, temporal pattern that permeates the object—rather than by any physical interaction among its parts. Even in the Copenhagen view, the trigger lies beyond the object: in Schrödinger's metaphor, it is the atomic decay that initiates everything.

Accordingly, the Schrödinger equation describes the system only up to the occurrence of a non-mechanical event, as defined in 2.4, which falls outside the scope of any purely mechanical picture. The consequence is clear: the Schrödinger equation—or, more generally, any mechanical description—must be abandoned at the moment of measurement. It neither explains nor can it explain what the external event (e.g., the measurement) actually is. All that remains is the propagation of possible events and their associated probabilities—as long as an ontology of phase is missing.

But the ontology is in fact evident. Because events and Evental Time are deeply intertwined, the phase of the wave function unmistakably encodes a temporal structure—one that points not merely to state-changes but to the very occurrence of process events. An isolated quantum system with perfect phase coherence ("no drift") evolves mechanically under the Schrödinger equation as long as it remains undisturbed. A process event is triggered precisely when the system undergoes a non-mechanical transition—most familiarly during a measurement interaction that lies outside its unitary Schrödinger evolution. Such an interaction establishes coherence between an external phase and the system's internal phase (recalling de Broglie's original idea), thereby selecting one outcome from the available possibilities via Born's rule, such that both phases effectively cancel. A useful representation of the external phase, developed in [3,4], is briefly discussed in Chapter 10. Conversely, any genuine inner–outer phase coherence counts as a process event: sometimes this leads to discrete transitions between energy levels—such as the emission of a photon—but it can also produce subtler, non-mechanical adjustments, such as the guiding-law shifts in Bohmian (average) trajectories or the pointer drifts observed in weak measurements, without immediately registering as an external outcome.

Furthermore, the inaccessibility of phase suggests that it is not fundamental within the framework of standard QM; rather, it points to a non-local process—one not yet fully understood within physics. Accordingly, we interpret the wave function as describing an evolution of evaluated time measures. What is traditionally called "collapse" now appears as a resonance—echoing de Broglie's idea once more—between internal and external time structures, evaluated with respect to the event that becomes actual.

8 Revision of the Axiomes of the Quantum Mechanics

Before commenting on the commonly accepted axioms (A1–A7), it should be noted that in the standard formulation the canonical postulate $[\boldsymbol{x},\boldsymbol{p}]=i\hbar$ is regarded as the underlying basis. Our historical perspective, culminating in Section 6.4, has shown that such a postulate is not required.

We therefore do not pursue the commutator here, since it merely enforces the Hilbert structure—a structure that we will later challenge in Chapter 10 by introducing global time correlations and making them accessible to experiment.

The standard formulation of QM rests on a few well-known postulates. For clarity, we list these axioms first before addressing them individually.

Overview of Commonly Accepted Postulates:

(A1) The wave function provides a complete description of physical reality.

(A2) Measurement causes the collapse of the wave function.

(A3) The Hamiltonian is a Hermitian operator.

(A4) Measurement outcomes correspond to eigenvalues of the Hamiltonian.

(A5) Superposition states are physically real.

(A6) QM is fundamentally probabilistic

(A7) The probabilities are calculated by Born's Rule

We now comment on each of these points in turn.

On A1:

This is not an axiom but a dogma. The claim of completeness is not genuinely empirical; it rests on the implicit assumption of the Markov property of nature, an assumption that has been challenged by Jacob Barandes [30,31]. Hidden variables are therefore not ruled out, since their dependence need not obey Markovian divisibility.

On A2:

This point has been repeatedly and forcefully challenged by Roger Penrose, particularly in his discussion with Curt Jaimungal (see [32]). He states: "Quantum theory as a whole is wrong. [...] Not Einstein is wrong—quantum mechanics is wrong. [...] Einstein and Schrödinger were much more polite. They said: 'It's just incomplete.' Incomplete means wrong."

Modern quantum optics and quantum information theory have moved beyond the axiom of instantaneous projection. Continuous measurement theory treats observation as a stochastic process evolving in Einsteinian time, described by stochastic Schrödinger equations and quantum trajectories (here the term is used extensively in the literature, but not in the classical sense—it denotes a formal representation of the evolving wave function), with the classical projection postulate recovered only in the limiting case of arbitrarily strong coupling (quantum Zeno effect, see [8],[9]).

While this development alleviates the harsh contrast between unitary dynamics and instantaneous collapse, it does not truly resolve it: the stochastic noise terms are introduced phenomenologically, and their ontological status remains unclear. Are they physically real, or do they merely encode our incomplete knowledge? The standard reply is to retreat to the density matrix as the supposedly "real" object, understood as an ensemble average over many trajectories. But such a density matrix is nothing more than a spacetime-distributed probability: the sharp collapse is not eliminated, only smeared across Hilbert space. This raises the question whether the replacement of sharp collapses by noisy ensemble averages is conceptually satisfying. A different situation would arise if one admitted trajectories in the sense of

David Bohm (see [33]), which could be seen as averaged paths derived from the phase structure itself—rather than as blurred probability distributions.

From our perspective, it remains to be clarified (see Chapter 10) whether the strategies investigated in [3,4] are also viable within the framework of QM. If so, then it can be shown that a stochastically differentiated treatment actually conceals a possible ontology rather than resolving it.

On A3:

As discussed in Section 5.3, the restriction to Hermitian operators is too narrow to capture the full phenomenology. Further developments in [18] and [2] provide an extended analysis of non-Hermitian QM, including time-asymmetric and open systems.

On A4:

Eigenvalues should not be seen as fundamental quantities in themselves, but rather as manifestations of coherent internal processes—such as the orbital dynamics of electrons in atoms, which underlie the ordered structure of atomic spectra. In Section 5.3, we showed that even in elementary generalizations of the Schrödinger equation, a dominant real drift term may replace a simple eigenvalue description. This drift, generally not reducible to a single number, characterizes the evolution. Even when a dominant real component exists—analogous to an eigenvalue—it typically does not correspond to a directly observable quantity. Rather, it governs the phase dynamics over time and becomes accessible only indirectly through interference effects or long-term structural features. Relevant examples are found in PT-symmetric QM (cf. [18]).

On A5:

The idea that superposition represents an actual physical property of objects is rooted in the assumption—explicit or implicit—that QM is either complete or must be supplemented by interpretive dogma. From our perspective, superposition merely reflects the openness of alternative pathways prior to an event (cf. Chapter 7). This calls for an expanded interpretation of physical reality. For instance, in a beamsplitter experiment, the fact that the wave function is split and later recombined does not imply that a particle literally 'takes both paths'. Rather, it reflects the inseparability of the experiment as a temporally structured, nonlocal entity. This view is reinforced by delayed-choice experiments, where merely allowing for which-path information—without actually using it—can erase interference patterns. Such phenomena imply that quantum experiments are not mechanistic processes governed solely by initial conditions, but holistic temporal structures. A change in the experimental setup—even after emission—modifies the outcome, because the wave function encodes the indivisible unity of internal and external time within the process.

On A6

The notion of an intrinsic probabilistic nature has been challenged in [3,4].

On A7:

ψ^* satisfies the conjugate Schrödinger equation, and both ψ and ψ^* yield identical probability values. This is often interpreted as implying that ψ^* represents a wave function propagating backward in time from the future. According to the results of Nelson [34] and Nagasawa [35], however, ψ and ψ^* emerge as two structurally distinct wave functions: ψ describes a forward diffusion process, while ψ^* corresponds to a backward (osmotic) pressure. Notably, Nagasawa's formulation is fully equivalent to the Schrödinger equation. This structural duality naturally points toward an underlying bilinear formulation (cf. [2]).

Taken together, these points reveal that the Copenhagen interpretation lacks any solid foundation: it rests on axioms originating in Heisenberg's approach (see 6.4), which—by truncating Fourier expansions of Newton's force law under the Ritz combination principle—had already excluded realistic trajectories, and which therefore cannot be regarded as genuine axioms derived from historical or empirical necessity. It was, in retrospect, a brilliant move to obtain a workable mathematics; yet at the same time it blocks any ontology, because it leaves unexplained what actually produces the Ritz combination principle.

At this point one may well ask: what, then, accounts for the Ritz combination principle itself? The striking fact is that Heisenberg invoked Newton's law of force, only to mutilate it by truncating its Fourier expansions in order to match Ritz. This step, while ingenious in securing a calculational scheme, left the underlying mechanism entirely unexplained.

Yet Heisenberg's theory is fully equivalent to Schrödinger's formulation, itself regarded as the basis of QM. Under critical scrutiny, the conventional axioms collapse, leaving the Copenhagen view without conceptual support.

The preceding discussion has shown that the commonly accepted axioms are neither immutable nor sufficient.

Historically, Heisenberg's procedure began from Newton's law of force, introduced a heuristic Fourier expansion, and then truncated it in accordance with Ritz's combination principle. This construction is mathematically equivalent to Schrödinger's formulation, which subsequently became the input for the philosophical interpretations of quantum mechanics. Within the Copenhagen school, the impossibility of resolving the quantum world any further was elevated to a principle. Yet such reasoning is circular: by truncating trajectories at the outset, their absence at the interpretative level was predetermined. This indicates, however, that the possibility of a sub-quantum description remains open.

To establish a firmer basis, we now restrict the axiomatic perspective to the very core of QM—atomic spectral theory—in order to clarify what the discipline is fundamentally about.

In order to clarify the foundations and in line with Heisenberg [27] (see Remark 6.4.1), the early quantum theory may be regarded as a theory of atomic spectra, grounded on a few classical and empirical pillars, summarized as follows:

1. Dynamics follow Newton's law of force.
2. Spectral lines obey Ritz's combination principle.
3. The wave phase assigned to the particle is determined by the Planck–Einstein relation applied to its relativistic energy.
4. The wave phase must remain Lorentz-invariant across inertial frames; this coherence condition is indispensable, since without it the ansatz collapses.

In the Einsteinian sense of special relativity, physical events (measured facts) are frame-independent: different observers describe the same set of spacetime loci where given conditions hold. Interference maxima and minima are such events; hence the coherence condition "phase = const." must be frame-independent. Consequently, the relevant phase—evaluated along the particle's worldline—is a Lorentz scalar. From this starting point, equations (6.2.1) and (6.2.2) (i.e. the Planck–Einstein relation and the de Broglie wavelength) follow directly.

Starting from Newton's law of force together with the relativity principle, the Heisenberg picture consistently follows from the experimental fact that spectral lines obey Ritz's combination principle. Since Schrödinger's wave mechanics is mathematically equivalent to the Heisenberg picture, the same conclusion applies to the standard formulation of quantum mechanics: classical mechanics, special relativity, and experiment already provide a sufficient foundation; no further assumptions are required.

In light of these considerations, the perspectives of Einstein and Schrödinger—both of whom maintained a realist position—appear well founded and motivated by the search for a deeper understanding rather than by adherence to conventional axioms.

Recognizing the wave phase as the relevant relativistic invariant marks the starting point for a revision of quantum mechanics within a broader understanding of time—an approach we pursue in the framework of Eventual Time.

Returning to quantum mechanics more generally, Dirac's transformation theory may be viewed as a natural extension; conversely, it still rests on this foundation. From this perspective, quantum mechanics is not an autonomous framework but rather a consequence of special relativity itself. Hence, attempts to "quantize" relativity appear misguided.

9 Quantum Mechanics as a Cross-Scale Theory of Time

9.1 Proper Time and the Phase of the Wave Function

The insights of semiclassical theory (WKBJ) suggest that the phase of the wave function should not be treated as a mere auxiliary quantity, but as a physical structure with the character of proper time. In relativity theory, proper time provides an invariant measure along a worldline, defined by the relation $S = -mc^2\,\tau$. The same structure appears in QM: any wave function can be expressed in polar form,

$$\psi = R\, e^{\frac{i}{\hbar}S}$$

where the phase S is the action-like variable. This representation, already introduced by Madelung and central to Bohm's interpretation [33], gains additional force when seen through semiclassical approximations. In WKBJ, the wave function locally reduces to an elementary wave with phase $\mathbf{p}\cdot\mathbf{x} - Et$, explicitly mirroring the plane-wave phase $p_\mu x^\mu$ of relativistic theory. EBK conditions then ensure that such phases close consistently across cycles, making coherence and discreteness inseparable.

From this perspective, the phase of the wave function can be regarded as the quantum analogue of proper time τ. Unlike in relativity, where τ is a fixed invariant, in QM it becomes a dynamical quantity inseparably tied to the system's evolution and is further influenced by an additional term derived from the wave function itself: Bohm's quantum potential [33]. In the WKBJ approximation the wave function locally reduces to an elementary wave with phase $\mathbf{p}\cdot\mathbf{x}$ - Et, explicitly mirroring the plane-wave phase $\boldsymbol{p}_\mu \boldsymbol{x}^\mu$ of relativistic theory. EBK conditions then ensure that such phases close consistently across cycles, linking coherence and discreteness. This identification, however, remains approximate: WKBJ works well in practice, but only locally, and corrections appear whenever amplitudes vary strongly or near turning points.

A polar decomposition of Schrödinger's equation, however, yields precisely the classical Hamilton–Jacobi equation supplemented by an additional term [33],

$$Q = -\frac{h^2}{2m}\frac{\nabla^2 R}{R},$$

which has the dimension of energy and is commonly referred to as the quantum potential. In this sense, Bohm correctly identified the missing contribution that turns the almost complete story into the full one. From our point of view, Bohm's trajectories should not be overemphasized; they can at most be regarded as averages over possible paths. What remains decisive is that Bohm's formulation reveals the additional structure that complements the semiclassical picture.

Taken together, the phase provides a time structure of the same formal status as proper time in relativity. The difference is that in relativity this structure describes a linear progression in a flat coordinate system, while in QM it is a cyclic quantity, driving circulation rather than straight motion. Our conclusion must therefore be read with these qualifications in mind: the kinship with relativity is established through the semiclassical approximation, while exact quantum dynamics introduces the additional contribution of the quantum potential. This does not weaken the argument but clarifies its scope: QM is not merely a probabilistic framework, but a time theory—closely related to relativity, yet extended by cyclic phases and their energetic costs.

9.2 The Cross-Scale Character of the Schrödinger Equation

We have pointed out elsewhere that Schrödinger's equation does not describe objects, but rather the relations between space, time, and energy. In this section, we examine the equation in more detail. An additional key point is that the Schrödinger equation (3.1) is inherently cross-scale.

The left-hand side, $i\hbar\frac{\partial\psi}{\partial t}$, converts energy into a phase-rotation rate; the right-hand side, $\hat{H}\psi$, encodes the energetic landscape (kinetic curvature plus potential). Without a conversion factor, these would remain incommensurable. It is precisely the pair $\hbar$ and $1/\hbar$ that equalizes scales, mapping energies to a common temporal rate so that from equation (3.5)

$$\frac{\partial\psi}{\partial t} = -\frac{i}{\hbar}\hat{H}\psi = i(\frac{\hbar}{2m}\nabla^2 - \frac{1}{\hbar}V)\psi,$$

i.e., curvature couples with the factor $\hbar$ (filtered, weak), while potential couples with $1/\hbar$ (direct, strong). This cross-scale mediation helps to explain why interference and nonlocal effects need not be seen as particle-like 'anomalies. The wave function binds scales: its phase transports temporal structure, and its amplitude encodes shape (curvature).

Consequently, QM is not confined to the microscopic; it is a scale-spanning theory whose temporal phase coherently links micro-, meso-, and macro-levels.

Empirical note. Time-selection experiments and mesoscale analyses (see OSF) are consistent with this view: the Schrödinger equation operates not merely as a computational device but as a structural bridge across temporal scales.

10 Empirical Probes - Time Selection and Temporal Coherence in Quantum Mechanics

In this chapter we introduce some elementary aspects of chronomorphy (see [3]) within QM, focusing on empirical probes of quantum time structures and their scale dependence. Chronomorphy denotes the shaping of physical phenomena through time: interference patterns and spectra themselves appear as forms of temporal structuring, and by applying *time selection*—i.e., conducting experiments only at pre-defined temporal relations (see Sec. 11.3 in [4])—such structures can be grouped, manipulated, or revealed. In a broader sense, as outlined in [3], the Fourier transform provides a universal translation tool: it maps any form into a time-carrying structure, in which the interplay of evaluated time measures (introduced in Section 2) can be isolated. By imposing restrictions or modifications and then retransforming, one can, much like in image processing, extract insights into the temporal structures inherent in a form. In the present context, however, we pursue a more elementary approach, using time selection to probe three fundamental domains:

- Interference patterns probe phase coherence,
- Spectral line shapes probe energy–phase correlations,
- Decay statistics test processes traditionally modeled as purely Poissonian.

The central hypothesis is that observed outcomes are not purely stochastic but can be systematically modulated by time choice. Correlated modulations across all three domains would point to a unifying temporal structure linking energy, phase, and timing within the framework of QM.

Rather than reconstructing full interference patterns, classifying atomic spectra in detail, or collecting exhaustive decay data, it may suffice to identify unusual events. The core idea is simple: through time selection, outcomes are classified only by whether they fall within the expected statistical frame or not. Experiments can then be performed at arbitrary times (as a reference) and at times determined by the selection rule. If unexpected events occur systematically more often under time selection, this would indicate that such "outliers" are not statistical artefacts but signatures of temporal structure.

10.1 General model (Interference, Spectra, and Time Trigger)

Let $q_n = Q(t_0)$ classify start times t_0 by time quality [3, Sec. 11.3 in [4]]. Under fixed conditions, each category yields an outcome K_n :

- I_n for interference,
- Σ_n for spectra,
- D_n for decay statistics.

The null hypothesis H_0: all K_n are drawn from the same distribution K_0.

Possible outcomes:

(i) No differences → H_0 holds, no time dependence.

(ii) Differences, no reconstruction of K_0 → additional structure beyond simple subensemble mixtures.

(iii) Differences with reconstruction of K_0 → time choice acts as a noncausal, correlative selector without new dynamics.

10.2 Case analysis: interference pattern

(i) No differences → H_0 holds.

(ii) (ii) Differences, no reconstruction → time choice correlates with outcomes; *average phase contributions* differ across time-selected ensembles, indicating hidden structure or incomplete state description.

(iii) Differences with reconstruction → time choice selects effective initial conditions; via Kocsis et al. [36, 37] one can test whether the implicitly selected trajectories align with the Bohmian velocity field.

Further remark. In our earlier work [3,4 in Sec. 11.3] we suggested that if the initial phase of the pulse is parameterized through specific time choice, then the elementary form $\phi_0 \sim \mathbf{p}_0 \mathbf{x}_0 / \hbar$ implies that a localized source ($\mathbf{x}_0$ fixed) corresponds to a spherical shell in momentum space. This opens the possibility of classifying ensembles of Bohmian trajectories according to the same temporal codings. While too specific for the present context, it indicates a natural extension of the present framework and suggests concrete avenues for testing.

Supplement: Part of the Heisenberg uncertainty relation, as emphasized by Bohm, resides in the choice of initial position; time choice makes this explicit, explaining the effective spread in position–momentum space.

10.3 Case analysis: Spectral Line Shapes

(i) No differences → H_0 holds.

(ii) Differences, no reconstruction → time choice correlates with spectral features; *phase correlations* differ between time-selected ensembles, revealing additional structure.

(iii) Differences with reconstruction → time choice implicitly selects initial conditions of the spectral process; observed broadening is then partly statistical rather than purely fundamental uncertainty.

10.4 Case analysis: Decay Statistics (Generalized Born's Rule)

The standard Born rule arises when the wave function is compared with itself, eliminating phases. The generalized rule (Eq. 2.3) compares two distinct wave functions ψ_1 and ψ_2. In our framework of time selection, this reflects the coexistence of two temporal structures: the process-local one, represented by the wave function ψ_2, and a global one, represented by ψ_1. An event occurs in this view at (x, t) only if $\langle\psi_1(x,t)|\psi_2(x,t)\rangle$ is real. Hence the corresponding phase condition sets bounds on the decay time delay as well.

We propose using a source of rare quantum events (mean interval 6–8 min, ($\lambda \approx (2.1\text{–}2.8) \times 10^{-3}\ s^{-1}$), realized via decay of a metastable state or a strongly attenuated Poisson source. On predefined time-selection days τ and random control days, a trial runs from a cold start until either the first event occurs or the fixed time window W expires. The primary variable is $p_\tau = Pr(T \leq W)$ or the decay rate Γ_τ; a significant deviation from control values implies a *phase-sensitive trigger probability.*

(i) No differences → H_0 holds.

(ii) Differences, no reconstruction → time choice correlates with triggering; *relative phases* differ across ensembles, beyond simple subensemble mixtures.

(iii) Differences with reconstruction → time choice acts as a noncausal selector modulating trigger probabilities without introducing new dynamics.

10.5 Practical considerations and data accumulation.

The time-choice/quality classification ([3,4]) using globally astronomically defined time structure has so far achieved high statistical significance only when limited to one well-controlled trial per day. Empirically, on the order of ~4,000 valid cases are required for a decisive test—so a naïve single-site implementation would span decades. A scalable remedy is to parallelize the experiment across multiple, ideally independent, locations. The individual p-values from those sites can then be combined (e.g., by Fisher's method, $-2\sum_i \ln p_i$, which follows a $\chi 2$ distribution with 2k degrees of freedom for k sites; see [38], p. 37) to increase power. To preserve the integrity of the correlative test one must:

- Run the experiments at carefully chosen sites that avoid mutual correlation via "Döbereiner coordinates," i.e., eliminate hidden common structure in the time-quality signal.
- Standardize design and execution (protocol, calibration, data formatting) so outcomes are commensurable and aggregatable.
- Keep organizational coupling loose: over-synchronization or tight coordination induces a spread in the effective time structure, analogous to decoherence/dispersion in quantum systems, which frustrates coherent accumulation of patterns.

Conclusion

We developed the concept of evaluated time measures and showed how it clarifies the role of complex numbers in quantum mechanics (QM). This perspective establishes a continuous line from de Broglie's phase waves to the modern Schrödinger equation, highlighting that quantum theory can be seen as a dynamics of time itself—an intrinsic temporal structure that underlies both QM and, as recent results suggest, relativistic dynamics as well.

Importantly, as can already be read off from the fundamental properties of quantum theory, this indicates a holistic concept of time. By its very nature, such a concept suggests a self-similar structure extending across micro-, meso-, and macroscopic domains. Consequently, the fundamental phase S_1 of the wave function could cohere with higher-order temporal structures S_n. Events would then be triggered precisely when S_1 and S_n are in coherence. In this sense, QM appears incomplete—because it captures only one projection of a potentially deeper temporal structure.

There is, however, a concrete prospect: if the experiments sketched in Chapter 10 succeed, it may become possible to complete QM in certain respects—locally, situationally, and with regard to specific questions. To achieve this, it is necessary to accept that time is neither an external coordinate nor merely a homogeneous backdrop, but rather a container of distinct qualities. By "qualities of time" we mean the selective emergence of patterns—such as interference structures—that arise through temporal choice. Unlike a uniform, unlayered model, time itself assumes a structuring function: it filters possibilities into concrete patterns.

In this sense, QM points beyond itself. Its present formalism encodes only part of the temporal structure of events. Future empirical tests of time selection may reveal whether a broader temporal framework can provide the missing link between phase, coherence, and the emergence of events.

In the ultimate analysis, QM is not merely a theory of quanta or chance, but a theory of temporal structure—concerned with the ordering of events. Relativity (RT), by contrast, is a theory of spacetime, grounded in the four-vector formalism and tied to the classical parameters of energy and momentum. To the extent that one interprets QM primarily as an event-oriented process, it may indeed be called a theory of time; but if one emphasizes the additional structures beyond apparent randomness, it is more appropriately viewed as a theory of spacetime, complementary to RT yet centered on events rather than trajectories. Therefore, attempts at a "theory of everything" may fail if RT and QM are unified primarily through quantization, rather than by recognizing their common temporal foundations.

As emphasized in Chapter 6 (see in particular Section 6.4), the true lineage of QM lies in classical and relativistic principles: Ritz's combination principle, the Planck–Einstein relation, and de Broglie's phase dynamics. The subsequent imposition of the commutator axiom narrowed this heritage to a single wave function, which is powerful for calculation but conceptually circular. It is circular because the very input—the Ritz combination principle—removes frequencies without explaining the mechanism of their suppression. As Definition 2.3 already indicates, and as the experiments proposed in Chapter 10 aim to test, precisely this missing explanatory step would point toward an ontology. Its omission, however, is cemented once the commutator is taken as fundamental.

This gap may be closed by recognizing that Ritz's principle itself points to a coherence relation—a structuring principle that manifests within atoms but is not confined to them. Outside bound systems, Newtonian mechanics suffices; inside, coherence reorganizes frequencies and produces spectral order. This distinction mirrors the difference between purely mechanical evolution and event-oriented structures: mechanics accounts for trajectories without events, while QM reveals coherence-triggered events as manifestations of a deeper temporal order.

Part II

Events and Time

A case study of the alignment of random processes with global temporal structures

Dr. Peter Paul Lehleiter

Independent Researcher and Publisher
Morellstr. 19
86159 Augsburg
Germany

Corresponding author:
Peter Paul Lehleiter
https://orcid.org/0009-0005-1997-8564
E-mail: Peter.Paul.Lehleiter@web.de

Contents

Acknowledgements

The author gratefully acknowledges the early inspiration and support of Dr. Elmar Hatzemann († January 2025), who conducted numerous excursions on the author's behalf and carefully recorded the data. Without his contribution, this research might never have started.

Abstract

This article proposes a new understanding of time: namely, as a manifestation of non-mechanical events. In contrast to the conventional view of clock time as a passive parameter, we argue that time actively shapes the formation of quantum events. The argument draws on the statistical evaluation of empirical data—specifically, series of roulette outcomes interpreted as time-ordered events under minimal anthropogenic influence. By uncovering subtle yet statistically significant correlations between observation times and astronomical structures, we demonstrate that temporal patterns emerge even in ostensibly random systems. These findings suggest that time is not a passive backdrop but a generative principle with measurable influence on event distributions.

Unlike clock time, which remains arbitrarily rescalable, *natural time* is fixed by astronomical cycles—a day is a day, its subdivisions anchored in planetary motion. Such non-scalable structures ground temporal order in real, observable events and establish a hierarchical link between global processes and local phenomena. In view of the proposed holoperspectival worldview, such cross-scale correlation is essential: at minimum, we require a model with empirical evidence that temporal structures consistently link global and local levels. This framework offers a principled basis for testing the empirical correlation of temporal qualities across different scales.

Our analysis employs non-parametric methods and high-resolution temporal coding, providing a solid foundation for rethinking measurement and causality. While this article focuses on the statistical and empirical aspects of the theory, it also lays the methodological groundwork for more speculative developments. These include a complexified time formalism and a reinterpretation of the Schrödinger equation in terms of time-structured dynamics, which will be explored in subsequent work.

Our analysis contrasts two complementary minimal approaches: a strictly probabilistic model of random outcomes (null hypothesis) and a globally defined concept of natural time derived from astronomical cycles. Demonstrating correlations between the two provides evidence for extending quantum mechanics toward a framework in which time functions as a real, structuring principle.

Keywords:
structured time, Eventual Time, event ontology, empirical quantum foundations

References

[1] P. P. Lehleiter, *Wie Zeit sich in Ereignissen manifestiert: Ansätze einer abstrakten Chronomorphie*. Augsburg: Chronomorph Verlag, 2024.

[2] P. P. Lehleiter, *The Temporal Structure of Quantum Events*. Augsburg: Chronomorph Verlag, 2025.

[3] S. M. Hoffmann, *Hipparchs Himmelsglobus: Ein Bindeglied in der babylonisch-griechischen Astrometrie?* Wiesbaden: Springer, 2017.

[4] S. M. Hoffmann, *Wie der Löwe an den Himmel kam – Auf den Spuren der Sternbilder*. Stuttgart: Franckh Kosmos, 2021.

[5] W. Döbereiner, *Astrologisch-Geographische Karten*. München: Verlag Döbereiner, 1993.

[6] W. Döbereiner and I. Bertrand, "Website," 2014. [Online]. Available: https://www.doebereiner-bertrand.de

[7] P. P. Lehleiter, "Geographical positions [Supplementary material]," Open Science Framework, 2025. [Online]. Available: https://osf.io/v2zy5

[8] D. Koch and A. Treindl, "Astro Dienst," 2014–24. [Online]. Available: http://www.astro.com

[9] W. Döbereiner, *Astrologisches Lehr- und Übungsbuch* (Vols. 1–6). München: Verlag Döbereiner, 1982.

[10] P. P. Lehleiter, "General number assignments [Supplementary material]," Open Science Framework, 2025. [Online]. Available: https://osf.io/7pgne (see *UsingOSFProject.pdf* in the root folder for project structure and file descriptions).

[11] P. P. Lehleiter, "Roulette Specific Number Assignments [Supplementary material]," Open Science Framework, 2025. [Online]. Available:

https://osf.io/7pgne (see *UsingOSFProject.pdf* in the root folder for project structure and file descriptions).

[12] J. Barandes, "The stochastic-quantum correspondence," *arXiv preprint* arXiv:2302.10778, 2023. [Online]. Available: https://arxiv.org/abs/2302.10778

1 Introduction: Time, Randomness, and Events

According to Einstein, time is what clocks measure. Yet this view remains circular, as it equates time with mechanical state changes. Our approach introduces the concept of *Eventual Time*: a structured temporal order that manifests in events themselves. Rather than relying on clock-based definitions, this concept enables the statistical examination of randomness for hidden temporal structures, which we refer to as *Implicate Time* (following David Bohm). In [1,2], the concept is elaborated in the context of quantum mechanics (QM), where it is proposed to interpret QM as a general principle across different scales. In this article, however, we focus on a simple random experiment at the mesoscopic scale (roulette outcomes) and demonstrate correlations with a temporal framework at the macroscopic scale (an astronomically defined time structure).

In this context, it is essential to emphasize why astronomical data are considered. Clock time, while precise, is arbitrarily rescalable and thus provides no natural anchor for experimental description. By contrast, astronomical cycles define *natural time*: a day is a day, its subdivisions fixed by planetary motion. Such non-scalable structures ground temporal order in real, observable events and establish a hierarchical link between global processes and local phenomena. This provides a principled basis for relating random outcomes to an objectively defined temporal framework.The advantage of this approach is twofold:

1. We provide empirical evidence supporting the arguments made in [1,2].
2. We show that models with hidden variables can also be realized at mesoscopic scales.

To test whether time emerges in dynamic systems—especially chaotic ones—we go beyond mere measurement. Instead, we assign specific astronomical objects numerical values, representing elementary time qualities. This framework is general enough for broad applicability while remaining empirically testable.

As an initial application of this idea, we present a statistically highly significant example based on empirical data. This concrete case marks an important first step. While many readers may find the argument compelling, some may remain skeptical given its implications. However, subsequent articles will provide further arguments and discuss additional ways to test the proposed framework.

The concept of *Eventual Time* emerged from the empirical observation that global temporal structures exhibit statistically significant correlations with locally stochastic processes. These correlations have been documented in detail in independent statistical analyses [1,2]. They demonstrate that what appears as pure randomness can, under certain global configurations, reveal residual structure.

It is essential to emphasize what *Eventual Time is not*. It does not rely on hidden forces or causal influences of celestial bodies, as sometimes claimed in astrological traditions. Nor does it invoke esoteric notions of fate or cosmic determinism. The analogy of a hologram illustrates this difference: the hologram does not contain atoms of the represented object, but it encodes structure such that, under appropriate illumination, the global form becomes visible. Similarly, Eventual Time does not postulate mysterious planetary influences, but recognizes correlations between global and local temporal orders.

Astrological interpretations claim causal power for planets and stars. In contrast, *Eventual Time* is a structural principle: local events can resonate with global time structures without invoking any physical force from celestial bodies. The distinction is crucial: this is not about *influences*, but about *correlations within a universal temporal order*.

Thus, Part B does not offer an esoteric speculation but a statistical demonstration: global temporal structures leave measurable traces in stochastic systems. This empirical grounding makes Eventual Time a candidate for integration with quantum theory, rather than a diversion into astrology.

2 The Incomplete Concept of Time in Physics—and a Straightforward Extension

2.1 Search for a Consistent Extension

From a physical perspective, the proposal fits into considerations of scales and controllability. Heisenberg's uncertainty relation, for example $\Delta x \Delta p \geq \hbar/2$, describes relations between mechanical variables. Quantum mechanics, however, is obviously not a mechanical theory: from the mainstream point of view, it is probabilistic and therefore non-mechanical. Within this framework, the wave function is globally defined in configuration space and, when conditioned by measurement, yields a specific local outcome according to the probabilistic rules of the theory. This effective selection—often heuristically referred to as "collapse"—demonstrates how a global structure can determine a single local value in accordance with quantum laws.

Collapse, however, should not be regarded as the final word of quantum mechanics. It is better understood as an interface between the established probabilistic formalism and a prospective extension. What manifests in probabilities is not an endpoint but a transition: the translation of global structures into local events. From this perspective, a single quantum event must already encode correlations across scales, even if these are not made explicit in the standard formalism.

Both quantum mechanics and relativity theory indicate that space and time are inseparable in a fundamental sense. What is therefore required is not another local observable, but a non-local spatio-temporal relation that can be meaningfully connected to the quantum framework. The task is thus to show that global temporal structures correlate with local processes in order to clarify the probabilistic description. The term correlation is used deliberately here, since by assumption no mechanical process can be adapted.

Since a direct demonstration is difficult to achieve, we choose a mesoscopic approach. This has the additional advantage of showing that quantum-like properties occur on multiple scales and phases. Their precondition is not a quantum-like "smallness" but rather an a-mechanical, probabilistic coupling.

In this way, a consistent extension of the physical framework becomes apparent: quantum mechanics, as argued in the first article, tells of a flow of evaluated temporal measures—yet the recognizable structure of these measures has so far not been taken into account. To prepare for this step, we next contrast two complementary minimal approaches.

2.2 Two Complementary Minimal Approaches to Temporal Structure

Here we deliberately present a coarse-grained starting point: On the one hand, there is a strictly local, purely probabilistic approach that formulates the null hypothesis—namely, that the outcomes of roulette games are purely random and carry no additional structure. On the other hand, we juxtapose a schematic, globally defined concept of time that is based solely on astronomical relations.

Both approaches are minimal and clearly delineated in themselves: the first reduces events to random outcomes that can be statistically described in their entirety.

The second relies on objectively determinable, non-scalable temporal structures—such as the day and its subdivisions, as given by planetary motion.

By setting these two extreme positions against each other, we establish the foundation for an empirical investigation: can correlations with globally defined temporal structures actually be detected in random outcomes? If so, this goes beyond a purely probabilistic model and opens the way to an understanding of time as a real, structuring principle.

Building on these minimal approaches, we now outline a broader conceptual framework that situates Eventual Time within a global structure

2.3 Conceptual Framework

First, we note: verification of the reality of time requires a fundamental departure from purely mechanical processes, in which time merely appears as an illusion. As Einstein explicitly stated—or is at least frequently cited as stating—a strictly mechanistic universe inevitably leads to this conclusion. So, there is no experimental access to the reality of time without a theory that allows time to be regarded as real. Therefore, we focus on complex and chaotic processes, which we regard as the source of events in our framework. We regard events as expressions of inherently unpredictable processes, through which time reveals itself as an active, structuring principle. Mechanical processes, by contrast, do not yield genuine events in this sense, but merely regular state changes. We consider events as the very manifestations through which time reveals itself, and their structure is, in principle, partially accessible through a simplified—yet global—temporal framework that contrasts with conventional notions of time.

Our conception of Eventual Time as globally structured is inspired by David Bohm's interpretation of quantum mechanics. Bohm emphasized that the unfolding of phenomena reflects a deeper implicate order—non-local, internally structured, and inherently temporal. Here, 'global' does not denote spatial extension but rather a universal temporal architecture underlying local events. Within this framework, even moderately large regions of the universe may exhibit holographic properties. Accordingly, we regard time itself as a holographic component of the universe. Such a representation preserves its essential informational content under coarse-graining or reduction, provided the segments are not too small. Thus, time can be legitimately treated as reduced in complexity yet enhanced relative to its environment, while retaining its structural significance for the present investigation.

Taken to its full consequence, we argue that process, event, and time cannot be meaningfully separated. A perfect generalized clock—what we would call a *Chronomorph*—would be identical to the process itself. However, such an identity has neither practical nor philosophical utility. Therefore, with the holographic principle in mind, we seek a simplified yet globally coherent temporal structure that makes essential features of *Eventual Time* empirically accessible.

This motivates a cautious approach to the idea that astronomical configurations—though often considered outdated—may still encode a relevant temporal structure. Within our framework, this is not a speculative claim, but a consequence of attributing intrinsic quality to time and understanding it as a holographic component of the universe. Under these premises, it becomes plausible that certain complex events reflect aspects of this structure—without invoking unverifiable mechanisms.

To avoid circular reasoning in the context of events, we refrain from imposing a strict classification of events. In particular, we do not distinguish between:

- Deterministic chaos, where randomness is merely regarded as an unexplained residual (pseudo-randomness), and
- Ideal randomness, where randomness is considered an intrinsic property of processes.

Our argument relies on two central theses:

T1

Time possesses a unique quality that cannot be captured by conventional physical parameters.

T2

Statements about random or pseudo-random events can be refined by employing an appropriate concept of time, extending beyond purely probabilistic considerations.

From our perspective, time exhibits a structured range of intrinsic qualities. Yet only certain aspects — such as those measurable by clocks or represented in calendars — become physically accessible. The particular quality that manifests in an event depends on the underlying process and its contextual embedding. The aim of this article is to discuss such elementary time qualities and to demonstrate their empirical relevance through statistical evaluation.

This becomes particularly evident in random experiments, where the observable outcome reflects a projection onto clock time (Einstein time: time as displayed clock time). The time quality itself is expressed in the result—the measurable aspect of the event—while the reason why this particular quality, and not another, emerges remains inaccessible within the framework of established physical parameters.

Our methodology consists of two steps:

1. **Development of a simple concept of time:**
   We formulate a time model based on astronomical relations that meets the criteria of generality and practicality.
2. **Experimental verification:**

Using a randomness experiment, we test the adequacy of this time concept and demonstrate its empirical relevance with respect to both T1 and T2.

Note:
All calculations can be verified using the supplementary material on OSF (https://osf.io/7pgne/files/osfstorage).

2.4 Natural Time and Astronomical Anchors

The motivation for employing astronomical data arises from the need to relate observed outcomes—manifestations of events—to time. Einsteinian time, as provided by clocks, is essentially a scalable parameter: its units can be arbitrarily chosen and therefore do not anchor experimental descriptions in a natural way. The question then arises: which temporal unit should be adopted? Our solution is to employ *natural time*, defined by astronomical cycles. A day is a day, not subject to rescaling; subdivisions of the day are likewise fixed. In this sense, astronomical time provides an intrinsic and non-scalable structure that allows events to be systematically related to temporal order.

In addition, the temporal structure is anchored in visible and real events. This is the decisive reason why such times are not scalable: their definition arises directly from observable processes. Beyond this, another essential aspect must be emphasized. Planetary motion itself constitutes a process that is hierarchically superordinate to local processes. Later, we will elaborate this within the holoperspectival principle, which provides a general scheme for relating an overarching process to a subordinate one.

The practical advantage is that the relevant data are both well-defined and readily accessible: on the one hand, as high-precision astronomical ephemerides (e.g.,

NASA data), and on the other hand, as documented sequences of random outcomes—casino permanences. In the case of the Wiesbaden Casino, as with some others, such outcome data are even recorded with second-level time stamps.The Well-Tempered Time—Alignment as a Principle of Wholeness

2.5 Motivation: The Universal Principle of Well-Tempering

We postulate a universal principle:

> Wholeness is not the result of perfectly separable individual components, but of their dynamic coherence. Conversely, in a truly holistic system, each part is shaped in relation to the whole.

A fitting analogy comes from music theory: when a piano is tuned to exact harmonic ratios, the sequences of octaves and fifths drift apart due to the Pythagorean comma. The well-tempered tuning of a piano deliberately introduces slight mistunings, enabling the instrument to function coherently across all keys. Likewise, we do not aim to capture Eventual Time in its full mathematical purity, but to develop a simplified, well-tempered structure that remains globally coherent and empirically usable. This principle can be applied to time measurement as well: rounding days and months creates an idealized structure that does not precisely match astronomical cycles, yet preserves intrinsic order and structural relevance [1].

Similarly, we adjust astronomical timeframes to exhibit a well-tempered structure. Just as twelve fifths are fitted into seven octaves, we align twelve new moons with the precise length of the solar year (365.2419) days. This points to a universal principle of well-tempering—manifesting wherever holistic structures operate, whether in music, time, or other complex systems.

With this approach, we demonstrate that time can be understood not merely as an abstract dimension, but as a real entity. The integration of well-tempered time structures opens new perspectives for studying randomness, chaotic processes, and their manifestation in events.

In subsequent parts of this article series, we will explore how complex numbers can be interpreted as intrinsic numbers of time, and how a generalized Born rule could provide a new foundation for analyzing quantum phenomena.

2.6 The Well-Tempered Calendar

2.6.1 The Well-Tempered Year

We define a well-tempered year as the time span from one vernal equinox to the next. It thus corresponds to the astronomical year—the time required for the Sun to complete one full orbit along the ecliptic (the apparent path of the Sun as seen from Earth), starting from the vernal equinox.

2.6.2 Well-Tempered Months and Days

The concepts of well-tempered months and days are based on dividing the ecliptic into equal segments:

- A well-tempered month is the time it takes for the Sun to traverse 1/12 of the ecliptic.
- A well-tempered day is the time required for the Sun to move 1/360 of the ecliptic.

Thus, an idealized well-tempered year consists of 12 well-tempered months and 360 well-tempered days, with each well-tempered month containing exactly 30 well-tempered days.

2.7 Differences in the Durations of Well-Tempered Months and Days

Well-tempered months and days do not have uniform durations. The abstract division of the ecliptic into equal segments of 30° per month and 1° per day does not imply that the sun traverses these segments in equal time intervals.

This variation arises due to the elliptical orbit of the Earth around the Sun, as described by Kepler's second law: The Earth moves faster when it is closer to the Sun (perihelion) and slower when it is farther away (aphelion). Consequently, the time required for the Sun to move through each segment of the ecliptic varies.

Therefore the well-tempered calendar integrates the dynamics of real, physical time and avoids rigid, uniform time intervals. This approach makes it possible to incorporate variations, such as years with only 11 or even 13 new moons, into a harmonized overall structure.

Analogous to music theory—where apparent "imperfections" allow for universal playability in all keys—the ideal concept of 12 new moons per year remains meaningful. This principle of adjustment is an integral part of well-tempering and serves as the foundation for a holistic understanding of time.

2.8 Well-Tempered Months and the Zodiac

The definition of well-tempered months can also be understood in a historical context: As early as the second century BCE, Hipparchus established the division of the ecliptic into 12 geometric sections of 30° each, forming the basis of what is now known as the zodiac (Hoffmann [3,4]). Consequently, the ecliptic is also referred to as the zodiac or the zodiacal belt.

However, it is important to distinguish between zodiac signs in Western astrology and actual star constellations. Zodiac signs are purely geometric divisions of the ecliptic, each spanning 30°. In contrast, star constellations refer to actual fixed star groupings, which are considered in Vedic astrology (Jyotisha).

This distinction highlights that the well-tempered time system remains independent of astronomical or culturally varying astrological interpretations, preserving its objective character.

3 Hierarchical Ordering of Objects

Having recognized that the solar path is useful for defining an abstract measure of time, we extend our consideration further. It is natural to also take other celestial bodies into account. The idea we pursue is to outline an astronomically defined concept of time, which serves as the basis for testing the previously mentioned theses T1 and T2.

3.1 The Moon plays a central role due to its:

- Brightness at night, which has made it a navigational reference point throughout history.
- Regularity of its phases, which has influenced timekeeping in the form of months.
- Influence on tides, caused by the interplay between the Moon and the Sun.

Given these factors, the Moon should be incorporated into our model alongside the Sun.

In addition to the Sun and Moon, the visible planets are also significant:

Mercury, Venus, Mars, Jupiter, and Saturn have been known since antiquity.

Later, with the invention of the telescope, additional planets were discovered:

- Uranus in 1781,
- Neptune in 1846,
- Pluto in 1930 (now classified as a dwarf planet).

These planets offer additional periodic phenomena that are of interest for defining astronomical time.

3.2 Solar and Lunar Eclipses

Solar and lunar eclipses have long been known and are impressive phenomena. They occur only when the Sun and Moon are near the intersections of their paths (the Moon's orbit and the ecliptic). These intersections are known as lunar nodes:

- The ascending node is the point where the Moon crosses from the southern to the northern hemisphere of the ecliptic.
- The descending node is the opposite point.

Eclipses can occur only when the Moon is near one of these nodes during the new or full moon phases.

Incorporating these celestial bodies and phenomena into our model allows for a more comprehensive understanding of astronomical time.

These considerations motivate the establishment of a hierarchy—termed the "natural order of objects"—which reflects the relevance of astronomical phenomena as structured appearances within time. Our aim is not to rank physical bodies per se, but to identify those configurations that contribute most meaningfully to the emergence of temporal structure. This sequence reflects the significance of astronomical entities within our model. We position the ascending lunar node at the final rank, as it is not a celestial body but an astronomically defined point on the ecliptic. The descending lunar node is not listed separately, given its opposition to the ascending node, and is thus implicitly included. Although the lunar nodes are astronomical constructs, they are considered within our enumeration of objects.

This layered system of astronomical references forms the basis for constructing a symbolic time code—a compact representation of time quality based on the relative positions of celestial bodies.

Alternative sequences could be considered:

1. Ordering by Brightness: Here, Mercury and Venus would switch positions, as Venus, alternately appearing as the morning and evening star, is significantly brighter than Mercury.
2. Ordering by Orbital Period: In this arrangement, objects are sorted strictly by their orbital periods along the ecliptic. The Moon would occupy the first position, with the Sun placed between Venus and Mars.

These alternative orderings offer different perspectives based on specific criteria, such as luminosity or orbital characteristics.

4 Houses and Cusps

From the perspective of a local observer, it would be natural to define the beginning of the ecliptic at the point where it rises above the horizon and to divide the observable sky into twelve segments. This viewpoint has indeed been formalized in mathematical terms. However, it suffers from an obvious limitation: our aim is to construct a global time structure—analogous to a calendar—that applies uniformly across the Earth, with only local adjustments (such as time zones). Therefore, we use the system of Placidus, dividing the ecliptic using actually progression phases. The twelve calculated division points are called *cusps*, and the segments between them are referred to as *houses*. We now seek a derivation of the Placidus house system grounded in our principle of well-tempering. We establish the temporal structure of the year by identifying four seasonal axes based on the progressive change in daylight: the Ascendant (AC), Medium Coeli (MC), Descendant (DC), and Imum Coeli (IC). These form the backbone of the house division and define four of the twelve cusps.

To complete the division, we adopt the Placidus system, which determines intermediate house cusps based on a measure-oriented division of the ecliptic. This method yields twelve segments—mostly unequal—reflecting the dynamic interplay between Earth's rotation and the observer's latitude.

To maintain coherence with the solar symbolism, we define Placidus degrees to increase clockwise: the AC is set to 0°, and the MC—analogous to the summer solstice—is set to 90°. The DC and IC follow at +180° relative to the AC and MC. For any given Placidus degree $p \in [0^\circ, 90^\circ]$, the corresponding position on the ecliptic is defined as the point g whose arc above the AC has risen to exactly p/90 of the quarter-circle from AC to MC. The other quadrants are treated analogously. The intermediate houses then correspond to those ecliptic positions where the Placidus degree is a multiple of 30°.

Although historically rooted in astrological tradition, the Placidus system gains renewed justification in our framework. It reflects an observer-specific, non-uniform segmentation of time that aligns with the well-tempering principle. Recent findings [1] support this model over other house systems, demonstrating that equal divisions based on the MC retain partial validity, while several commonly used alternatives fail empirical scrutiny.

Traditionally, houses are numbered counterclockwise. However, since Placidus degrees are measured clockwise, the 12th house begins at 30° Placidus, and the 11th house at 60° Placidus. The remaining houses are determined by shifting their positions on the ecliptic by 180° accordingly.

5 Geographical Positions and Coordinates according to Wolfgang Döbereiner

The house division provides only a rough framework and lacks the spatial resolution needed to pinpoint local effects in time-structure experiments. In mid-latitudes, the area where both the Ascendant (AC) and Medium Coeli (MC) deviate by less than one degree still spans several hundred square kilometers—underscoring the need for finer localization.

This is where the coordinate grids developed by Wolfgang Döbereiner come into play. First published in a loose-leaf collection [5], they were later made available as software in collaboration with Imanuel Bertrand [6]. Based on the exact geographical positions [7] we derived, we determined precise coordinates according to Döbereiner using [6].

Döbereiner distinguishes between two grids: the **macro-coordinates** (Großkoordinaten) and the **micro-coordinates** (Kleinkoordinaten). Both grids define a **oblique line** (*Schräge*) and a **transverse line** (*Quere*) for each geographical location:

- The **oblique line** is a loxodrome inclined at 23° to the meridian and running in a north-northwest direction.
- The **transverse line** runs parallel to a latitude.

In mid-latitudes, the macro-coordinates have a spacing of approximately 15 kilometers per degree (on zodiac). In contrast, the micro-coordinates allow for a significantly finer subdivision—about 50 meters per degree. This refined grid enables the accurate specification of geographical positions and is essential for integrating experimental results into a spatially resolved time structure. It bridges the gap between the global, calendar-like time framework and the local variations relevant to empirical analysis.

In the following sections, we examine game results from the Casino Wiesbaden. To ensure sufficient precision—particularly for the application of micro-coordinates—geographical positions must be determined with meter-level accuracy. These positions are listed in Supplementary Table 1 [7].

The non-smoking section is no longer in operation today. However, since the author archived the data over time, it is still included in the analysis. Additional analyses showed that excluding the days on which the former non-smoking table initiated the daily games has a negligible impact on the results.

The numbered tables represent traditional roulette tables, while those labeled with a short text refer to Touch-Bet-Terminals. In these machines, the ball is launched using compressed air, and bets are placed via a touch screen terminal.

The houses and object positions were calculated using the Swiss Ephemeris [8], while the Döbereiner coordinates were determined using Horoskop 5.0, see [1].

We denote the macro- and micro-coordinates according to Döbereiner as WDGK ("Großkoordinaten") and WDKK ("Kleinkoordinaten"), respectively, including their difference and sum relations. Accordingly, WDXY represents all Döbereiner coordinates. Table 1 below shows the maximum difference in Döbereiner coordinates for the respective game options considered. Here, "S" stands for Schräge (oblique line) and "Q" for Quere (transverse line).

Table 1: Maximum coordinate deviation

WDGK		WDKK	
WDGKS	WDGKQ	WDKKS	WDKKQ
2''	0''	2.8°	0.8°

For the Döbereiner macro-coordinates, we observe a maximum deviation of two minutes among the considered tables. Since the houses for the different locations we determined for the tables are almost identical, using the Döbereiner macro-coordinates practically results in no differentiation based on location.

For the micro-coordinates, however, we find a maximum deviation of 0.8° for the transverse line (*Quere*) and even 2.8° for the oblique line (*Schräge*). These are significant differences.

6 The Search Algorithm for Relevant Relations in the Train Station Model (TSM)

In this metaphor, the train station represents a hierarchical structure. When standing at platform 1, if a train is present, you can only see the foremost train; other carriages behind it or trains on other platforms are not fully visible. This imagery illustrates how, within a hierarchy, higher-level components can obscure or overshadow those at lower levels, emphasizing the prominence of the leading element and the hidden nature of those that follow.

For the analysis, we calculate the difference relation and the sum relation for each pair of objects—including houses and Döbereiner coordinates as objects. These are derived from the difference or sum of their respective positions, normalized to the range of 0° to 360° (exclusive). If no explicit distinction is made between sum or difference, we generally refer to them as relations. If P_1 and P_2 are objects or their positions, we denote a relation R between them as $R = P_1 \sim P_2$. Here ' ~' can represent either ' +' or ' -'.

In the computed relations, we observe a recurrence of the proportional relationships found in the zodiac. We pay special attention to values that are approximately multiples of 30°. Square aspects—i.e., the angles of 0°, 90°, 180°, and 270°, which we treat together for simplicity—are assigned even greater significance.

A relation remains valid even if the ideal values are not met exactly. The allowed deviation—referred to as the *orb*—must not exceed the specified tolerance value.

To better represent the structure, we introduce the concept of the *base theil*: its specification implicitly includes all integer multiples, which we refer to as *theil values*.

Note: The term *theil* is derived from the Middle High German *teil* (also spelled *theil*), meaning both "part" and "period of time." In our model, it denotes structured temporal-angular divisions.

These adjustments aim to enhance clarity and ensure consistency with commonly used terminology.

The following Table 2 summarizes how the relations are considered in our analysis:

Table 2: Theil and Tolerance Values for Relevant Relations

Relation Partner 1	Relation Partner 2	Base Theil	Tolerance Value
Object 1	Cusp	90	1.2
Object 1	Döbereiner Coordinate	90	1.2
Object 1	Object 2	30	3.2

Definition 7.1.

Let P_1 and P_2 be objects (e.g., Sun, Moon, planets, or lunar nodes), and let R be a relation defined as $R = P_1 \sim P_2$, with a base theil of 30° and a tolerance (*orb*) of 3.2°. The relation R is called *relevant* if, for either $Q = P_1$ or $Q = P_2$, the following two conditions **are both satisfied**—each using a base theil of 90° and an orb of 1.2°:

- Q has a relation to a house.
- Q has a relation to a Döbereiner coordinate.

Thus, for $R = P_1 \sim P_2$ to be considered relevant, both criteria must be fulfilled by either P_1 or P_2. Relaxing this requirement—e.g., allowing P_1 to fulfill one condition and P_2 the other—has consistently led to significantly worse results. These weaker variants are not further discussed in the results section to avoid overloading the text with tables and graphs.

Accordingly, all relations that are integer multiples of 30° are included in the analysis. This also includes mirror points (*Spiegelpunkte* in the terminology of Döbereiner, who restricts them to squares [3,4]), which are treated as sum relations and constitute a special case within this general structure.

Our approach is structured as follows:

1. **The search algorithm:** Searching for a relevant relation using the natural sequence of objects.

2. **Defining a strategy** based on the relevant relation found in step 1 and the tables of roulette-specific number assignments.

We now define the TSM by establishing a search algorithm for a (unique) relevant relation. Reference [1] also examines the "Merge" model, which, by considering all relevant relations, still produces good results. Readers interested in further details can refer to that work.

Search Algorithm for a Relevant Relation $R = P_1 \sim P_2$

The goal of the search is to identify a *relevant relation* $R = P_1 \sim P_2$, based on the criteria defined above. The procedure follows a hierarchical loop structure, iterating through candidate objects, base theils, and tolerance levels.

1. Select the base theil: Start with 90°,

    a. Set the initial tolerance value: Begin with 1.2°, increasing by 1° per iteration, up to a maximum of 3.2° (three steps for this base theil).

        (1) Select the first object P_1, starting with the Sun and proceeding through the natural sequence of objects.

        (2) Select the second object P_2, beginning with the Moon and continuing likewise through the sequence.

        (3) Check whether $R = P_1 \sim P_2$ qualifies as a relevant relation.

        (4) If so, terminate the search.

        (5) If not, proceed to the next combination in the loop.

2. Proceed to 30° and carry out the above points in 1. a.,(1)-(5).

Once a relevant relation has been found, a corresponding strategy is defined (see section *The Train Station Model (TSM) and Hypothetical Betting*), and the outcome is evaluated accordingly.

Definition 7.2.
A relevant relation is called *essential* if it is returned as a solution by the search algorithm above.

7 Number Assignments out of Experience

In the previous section, we used the search algorithm to identify relevant relations. The next step is to assign a symbolic quality to these relations—achieved in the simplest way by assigning numbers to them.

Many decades ago, Wolfgang Döbereiner critically examined traditional number assignments through observational studies conducted during casino visits. He concluded that—with the exception of Saturn, correctly assigned the number 4—all traditional assignments were incorrect. His revised values (see [9] and Table 1 in [10]) form the basis of our approach, which I have subsequently extended based on my own empirical findings.

In addition to the fundamental number assignments for individual objects, several relation-based assignments were also derived empirically. The complete tables are provided in Supplementary Material Table 2 [10].

Note:

- Aggregated relation numbers—formed by concatenating individual assignments (e.g., 3 and 4 → 34)—were tested but found to have no meaningful effect in the analysis of permanences at the Casino Wiesbaden. Therefore, they are not considered here. For further details, see [1].
- Using only the suggestive Final Numbers yields a p-value of ≈ 0.0041.Roulette-Specific Number Assignments.

7.1 Macro- and Microstructure

The number assignments presented so far have been developed independently of the specific configuration of the observed system. We now shift our focus to local structures that may be relevant for the analysis of permanences at the Casino Wiesbaden. Two structural levels are distinguished:

1. **Macrostructure**
   Arrangement and number of objects at a location (e.g., position and number of roulette tables in a casino hall).

2. **Microstructure**
   Characteristics of individual objects (e.g., design of the roulette wheels).

Example:

The spatial arrangement of tables could theoretically influence results. However, a stronger influence is more likely to stem from the microstructure—especially the roulette wheel itself. We therefore simplify:

- Macrostructural differences are ignored.
- The focus lies on the microstructure of the roulette wheels.
- Differences in wheel size, layout, functionality, or occupancy are not considered.

Note:

During early experiments at home, I developed initial principles for number assignments, which were later refined for greater systematic coherence. These refinements did not materially alter the statistical results; the significance only decreased slightly.

Our analysis focuses exclusively on French Roulette.

Our analysis focuses exclusively on French Roulette. We assume a fictional payout of 37 units for a win on a single number—*including the stake*, thus ensuring that the expected value of the game is zero. However, the actual payout is only 36 units, resulting in a negative expected value of -2.70%.

We provide a brief overview of how roulette-specific numbers were derived. The referenced tables are available in [1,10]. A full description of their role, along with all relevant tables—including the roulette-specific numbers—can be found in [1].

7.2 Categories of Number Assignments

Number assignments fall into two main categories:

1. **General Number Assignments**
   - *Elementary Number Assignments* [1,10]
     1. Based on Döbereiner (*Döbereiner Singles*)
     2. Based on own experience
   - *Aggregated Numbers (as mentioned above, not taken into account)*
   - *Empirical Relation Numbers* [1,10]
     1. According to Döbereiner
     2. Out of own experience
2. **Roulette-Specific Number Assignments**

Based on structural features (Microstructure) of the roulette wheel [1].

7.3 Roulette-Specific Number Assignments

These assignments integrate both general and roulette-specific logic, grounded in a strict structural approach. While the general assignments are based on Döbereiner's work and my own empirical observations, the roulette-specific ones incorporate the physical layout of the wheel.

The multitude of rules used to construct the number assignments may be confusing to some readers. Certain rules might be questioned or interpreted differently. However, corresponding analyses have shown that slight modifications to the rules result in only minor changes in statistical significance.

To spare the reader such considerations, an additional analysis was conducted using only the rule of final neighbors. Naturally, this has consequences for the significance level: since each rule contributes to the fictitious final gain, the p-value tends to decrease rapidly as additional rules lead to further gains.

It is important to note that only 32 out of 55 possible relations received such number assignments. Since missing assignments are usually completed using Döbereiner numbers, this fallback was deactivated in the present analysis, as it would otherwise apply to 23 of the possible relations.

> **Note:** Even when using only the final neighbors, a respectable significance level *well below one percent* is still achieved. The full set of rules, however, improves the significance by approximately three orders of magnitude.

7.3.1 Final Neighbors

Final neighbors are numbers whose final digits match the Döbereiner Singles of two related objects and who are adjacent on the roulette wheel. *Example:* Venus = 5, Moon Node = 2 → 15 and 32 are neighbors on the wheel and end in 5 and 2 → 15 and 32 are *final neighbors* for the relation Venus ~ Moon Node.

Only two-digit numbers are considered. Single-digit Döbereiner numbers and supplementary numbers are not used as final neighbors themselves but may serve as endpoints.

7.3.2 Gap Numbers

Gap numbers lie between two Döbereiner numbers on the wheel and represent the relation between them. *Example:* 23 lies between 8 (Pluto) and 10 (Sun) → gap number for Sun ~ Pluto.

7.3.3 Gap Numbers of Type 0

In *Döbereiner missing sectors* (e.g., 26–0–32–15–19...25–34...24–31...28–35), no Döbereiner numbers occur. To balance these regions, *Gap Numbers of Type 0* are introduced: numbers that lie between a Döbereiner number and a supplementary number, or between two supplementary numbers.

7.3.4 Gap Numbers of Type 1

These lie between two numbers assigned to the same relation. Permissible endpoints include final neighbors, aggregations, or supplementary numbers. *Example:* 18 lies between 22 and 29 → both assigned to Mars ~ Moon Node → 18 is a *Gap Number 1.*

A one-one-bet covers a number together with its two immediate neighbors on the roulette wheel (a 3-number neighbor set).

7.3.5 One-One Gap Numbers

These are numbers that lie between two such one-one-bets. Example: Number 12 lies between the sets 3–26–35 and 29–7–28 and is therefore classified as a one-one gap number—provided it falls within a missing sector.

7.3.6 One-One Gap Numbers of Type (1)

Extension of the above: endpoints may include Döbereiner or supplementary numbers. The number in between is still required to lie within a missing sector.

7.3.7 One-One Final Supplements

These are overnext neighbors of a Döbereiner number with a matching final digit, used only if they lie in a missing sector. *Example:* For the relation Venus ~

Mercury, the set 5–6–10–24–27–34 may miss 16, which ends in 6 and is a neighbor of 24 → 16 is a *one-one final supplement*.

7.3.8 Final-Gap Numbers

These combine the logic of final neighbors and gap numbers. *Example:* Between 2 (Moon Node) and 17 (ending in 7, Moon) lies 25 → 25 is a *Final-Gap Number* for Moon ~ Moon Node.

Only gaps between at least one single-digit Döbereiner number and another qualifying number are considered. Two-digit-to-two-digit gaps are excluded to avoid inflation of the category.

Special care is required in constructing roulette-specific number assignments, especially in *Döbereiner missing sectors* where direct mapping is not possible. To maintain structural balance:

1. **fundamental number assignments** (according to Döbereiner) **must not be reused as gaps or neighbors.**

2. **Numbers from Döbereiner missing sectors must be included via special rules**, such as *Gap 0* categories.

See [1] for details and examples. For the full list of number assignments see [11].

8 The Train Station Model (TSM) and Hypothetical Betting

8.1 Hypothetical Betting

To deviate from actual casino conditions, we assume a fictional payout of 37 units *including the stake* for a win on a single number—ensuring that the expected value of the game is zero. This reflects the fact that the wager could have been placed on any of the 37 numbers (0–36).

A strategy is defined by first identifying a relevant relation via the search algorithm described above. Once found, number assignments are compiled according to the rules outlined previously or as detailed in the section *Roulette-Specific Number Assignments* in [1]. If no roulette-specific entries exist for a relation $R = P_1 \sim P_2$, the Döbereiner Singles are used.

Using this strategy, a hypothetical game is simulated. Bets are placed on the assigned numbers using a fixed chip value of 100 monetary units. Depending on the relation's number table, multiple chips may be assigned to the same number.

8.2 Used Coups and Definition of the Hypothetical Result

The observation period spans from October 1, 2003, to March 24, 2023 (7,115 days). For each day, only the *first coups* from the *opening table*—the table that began play first—are considered. It is not verified whether another table began at the exact same time, as such coincidences are rare at the Casino Wiesbaden.

Touch-Bet-Terminals are treated as tables.

For each point in time, the *hypothetical result* ("result" for short) is computed as the cumulative balance of hypothetical winnings minus bets. The relative *result* is the percentage ratio of this fictional result to the total hypothetical bet sum up to that point.

Days with missing or corrupted permanence data are excluded. A list of these dates is provided in Appendix A3 of [1]. However, some days with implausible data (e.g., only one recorded result) were retained, as their inclusion does not affect the null hypothesis of randomness. Therefore, no data correction was applied for these occurrences.

9 Results for the TSM

Note: *The results are documented in detail in the Testing Workbook on OSF* (https://osf.io/7pgne/files/osfstorage).

The relative results for the TSM are shown in Figure 1. Each point represents the ratio of cumulative fictional gains to total bets placed up to that year.

The first 1,000 data entries were omitted, as early results (especially within the first three years) exhibit extreme outliers—up to–100%—which would distort the visual scale and reduce clarity.

Figure 2 below compares the developments for:

- All Döbereiner coordinates (WDXY),
- Only major coordinates (WDGK),
- Only minor coordinates (WDKK).

While the three curves converge in later years, they differ markedly in the early phase. The reason for this divergence remains unclear. Notably, the WDGK trajectory rises only slightly between 2012 and 2020, whereas WDKK shows a more consistent upward trend.

This is especially striking given that the WDGK values for the various tables differ only marginally. Yet, by the end of the observation period, both curves (WDGK and WDKK) reach their overall maximum or settle just below it.

The results are evaluated using:

- **Linear regression**, and
- **Monte Carlo simulation**: Using Python's Mersenne Twister pseudorandom number generator, random sequences were generated to simulate game outcomes. The number of simulations yielding results equal to or better than the observed values is denoted *n*, with *N* total simulations. The estimated error probability is then $p = n/N$.

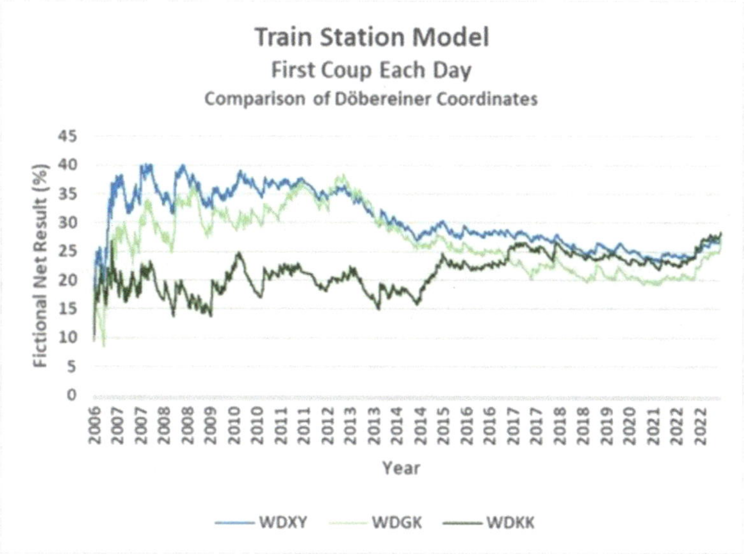

Figure 1: Relative Results for the TSM

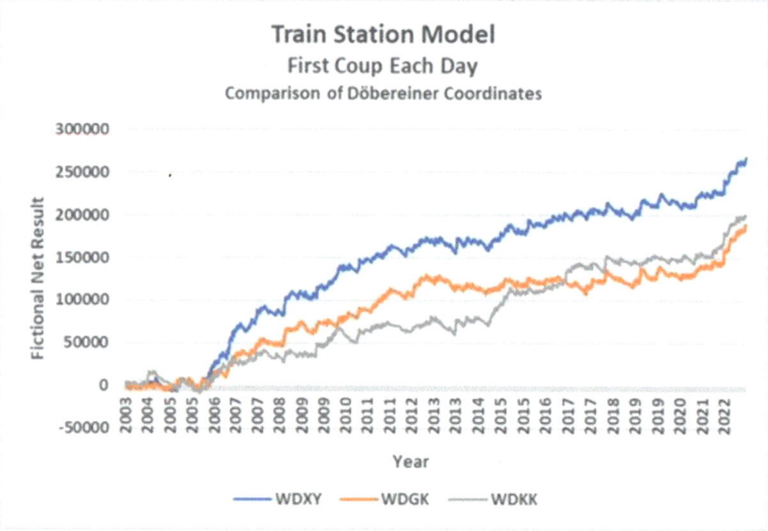

Figure 2: Net Results for the TSM

9.1 Statistical Analysis

Table 3 summarizes key statistical indicators:

- Coefficient of determination (R^2)
- Error probability (p-value).

The p-value was determined with high precision based on one million Monte Carlo simulations. As it reflects the probability that the observed fictional net result arose by chance, the outcome can be considered statistically significant.

Table 4 presents the most important figures in compact form. For comparability, the last row shows the simulated p-value and the corresponding multiple of the standard deviation ($\sigma = 1$), as derived from the normal distribution.

Table 3: Statistical summary of net wins, σ-estimates, and significance levels (Döbereiner coordinates)

	Döbereiner coordinates		
	WDXY	WDGK	WDKK
Total fictional net wins	266,800	188,800	200,900
Estimated standard deviation (σ)	52,252.42	52,955.22	52,227.67
Empirical σ-factor	5.1	3.6	3.8
p-Value	1.0×10^{-5}	3.3×10^{-4}	1.1×10^{-4}
σ-factor (Gaussian reference)	4.3	3.4	3.7

9.2 Significance and Interpretation

In particle physics, a significance level of 5σ is considered the gold standard due to extensive, computer-assisted analyses and repeated hypothesis testing, which increase the chance of false positives.

This problem does not apply here. The roulette-specific numbers were developed manually on a home roulette wheel and only later tested against real data from the Casino Wiesbaden. Earlier studies focused solely on fixed object-to-house assignments or exact relation values (e.g., integers).

In contrast, the present machine-based analysis for the first time includes relations with variable orb values within defined tolerance limits and integrates Döbereiner coordinates, which had not been part of earlier automated investigations.

Conclusion: Even a less extreme p-value here can be considered comparable in evidential strength to the 5σ threshold used at CERN. Table 4 presents the statistical key

figures, including results from linear regression:

- R^2 values range from 0.868 to 0.961, indicating strong trend stability.
- The slope of the regression line (growth rate of the fictional result per game) lies between 21.6% and 25.8%, relative to average stake size.

- The intercept remains proportionally small, supporting model robustness.
- The betting amounts show no trend, ruling out a hidden influence on the net result.

In particular, Döbereiner minor coordinates (WDKK) show more stable and specific effects than the more general major coordinates (WDGK)—reflected in higher average profits, slopes, and R^2 values.

A comparison between Casino Wiesbaden and the city center of Mainz shows a difference of only –8' in the Ascendant at the same time. In Central Europe, about 15 km are needed to shift Döbereiner inclinations by 1°. Thus, major coordinate values for different tables or terminals in Wiesbaden differ only marginally. Consequently, the TSM result does not depend on individual tables, as all are within a few dozen meters.

Interpretation:

A single casino might behave as if it were entangled with all others—provided relevant relations are initialized under similar local time-space conditions. However, this entanglement dissolves over larger distances, as major coordinates diverge rapidly with location and opening hours are fixed.

Note: The dataset includes archived data from the former *non-smoker Touch-Bet-Terminal*, which is no longer publicly accessible. Relevant exclusions had only a marginal effect and are therefore not detailed further.

In the variant using only the final neighbors (without supplementing missing assignments), the fictitious total gain amounts to *110,700*, corresponding to an error probability of $p \approx 0.0041$, based on 1,000,000 simulations. This shows that even when the set of rules is condensed to a few easily understandable ones, statistical significance is still preserved.

Table 4: Results of the TSM with 1,000,000 Simulations (in 1,000 units)

	WDXY	WDGK	WDKK
Total bet amount	983,8	736,2	720,4
Count of fictional bets	4,076	3,065	3,029
Average bet size	241.36	240.20	237.83
Total fictional net wins	266,800	188,800	200,900
Minimum	-7	-5,3	-9,2
Maximum	267,3	189,3	201,2
Average win per single bet	65.46	61.60	66.33
Average wins (%)	27.1%	25.6%	27.9%
Intercept	14,916.92	10,488.35	13,515.71
Slope	60.95	51.99	61.45
R^2	0.929	0.868	0.961
p-Value	1.0×10^{-5}	3.3×10^{-4}	1.1×10^{-4}

In addition, deviations from the standard number assignments—specifically, by omitting the selective use of Döbereiner numbers in cases where no structural rule applies—lead to a nearly tenfold increase in the error probability. This highlights that the semantic assignments are not arbitrarily additive, but rather serve as targeted complements where structural patterns alone remain incomplete.

Additional analyses in [1] highlight the central role of Döbereiner coordinates:

- Replacing these coordinates with random values—while leaving all other parameters unchanged—results in purely random patterns.
- If both the coordinates and the house system are randomized, the outcome resembles a classic random walk.

Table 4 and its discussion support both Hypothesis T2 and, indirectly, T1. This structured assignment process gives the relations a defined quality, reflected in their numerical representation. Moreover, in [1], an additional simulation was conducted: for each data record, a random strategy was generated by

drawing a random number between 0 and 36, using the previously computed strategy as a probability distribution (i.e., the assigned bets on individual numbers served as probability weights), and

- placing a fictional bet of 100 monetary units on the selected number.
- Despite this simplification, the variant still yielded a significance level of less than 0.01.

This further supports the thesis that the strategies contain non-random winning components and represent temporal qualities in a homogeneous and structured manner.

10 Conclusion

Standard quantum mechanics overlooks a crucial aspect of nonlocality. It usually investigates correlations by placing two detectors far apart. In entanglement experiments, nonlocality is defined through the spatial separation of distant detectors.

Our investigations (see [1]), however, focused exclusively on relations and did not take into account the actual distances of the involved objects. These distances can differ substantially—for instance, Pluto's mean distance from Earth is about 5.9 billion kilometers and varies by millions of kilometers, without affecting the results. Among the planets (with Sun and Moon not considered as planets here), Pluto shows the strongest evidence. This suggests that the underlying time structure is global and not bound to spatial locality.

By contrast, the Train Station Model (TSM) incorporates astronomical data and points to a radically different kind of nonlocality—holographic in nature. In this view, any process at any location may, in principle, reflect the entire structure of its environment, even its most distant components.

Thus, the holographic property of the universe, and its nonlocality with respect to time, has so far been overlooked in quantum mechanics. Therefore, rather than speaking merely of a holographic universe, we adopt the broader term *Holoperspectival Universe*, emphasizing that all perspectives are embedded in and shaped by a unified temporal-structural whole.

In our analysis, we examined the relation between ongoing ecliptic positions and chaotic processes. Using simple groupings, we were able to assign statistically significant qualitative properties to time—though this represents only an initial indication.

Following Bohm's terminology, Eventual Time may function as a container of enfolded structures that later unfold at specific Einsteinian times. In our context, this suggests that chaotic processes may in general depend on past ecliptic configurations.

Consequently, a Markovian model of the universe—as currently dominant in physics—appears inadequate. Such a framework extends beyond spatial relations to include temporal ones—at least in retrospective terms. These considerations do not constitute a direct critique of existing entanglement experiments. Rather, they suggest that the theoretical frameworks and experimental setups used in such studies may require further development and reinterpretation. From the present perspective, it appears increasingly plausible that the assumptions underlying these experiments—particularly the Markovian character of the universe—should be re-evaluated in light of temporally structured effects.

We will further develop these ideas in the subsequent article. At the same time, this perspective removes a potential objection to Jacob Barandes' approach: his concept of indivisible probability offers a viable framework precisely because it does not rely on Markovianity.

Our investigation has demonstrated that structured number assignments—rooted in both astronomical alignments and empirical observations—can provide meaningful insight into seemingly random events such as roulette outcomes. The Train Station Model (TSM) has proven effective in identifying relevant relations, particularly through a systematic approach that incorporates Döbereiner numbers and phase-based structures derived from the zodiac. These findings challenge the traditional assumption that random events lack deeper structural patterns, offering a fresh perspective on probability and time-dependent phenomena.

10.1 From Abstract Structural Components to Ontological Structures: Interpreting the Complex Phases in Barandes' Θ-Matrix

Building on the matrix-based formalism proposed by Jacob Barandes [12], we introduce an analogous decomposition $\Theta_{ij}(t) = r_{ij}(t) \cdot \omega_{ij}(t)$, where the phase term captures structured temporal modulation and the pair (i, j) not as abstract states, but as consecutive roulette outcomes (coups).

The time-dependent transition probability $r_{ij}(t)$ is defined as the probability that, given a coup i we observed at time t, next coup j. Remember that we only observe the first game result each day. As the first game at the Casino Wiesbaden does not begin exactly at the planned opening time (differs up to some minutes), the probability mass of $r_{ij}(\cdot)$ is concentrated on the first minutes after opening each day.

Moreover, if the relation i~j is relevant and object j moves slowly, it is slightly more likely that the same relation remains relevant the following day, especially around similar times.

This introduces a weak but statistically significant dependence in the sequence of daily outcomes.

Because the TSM yields a significance level well below 1.0×10^{-5}, these dependencies are not artifacts but reflect an underlying structure.

In this sense, $r_{ij}(t)$ plays a role analogous to the time-dependent weights in Barandes' formulation, capturing a real but nontrivial temporal stochastic structure.

The structural factor $\Theta_{ij}(t)$ encodes how well the relation i~j aligns with the hierarchical time framework (object-object, object-house, and object-coordinate relations).

The angular component evolves continuously along the ecliptic and defines a rotation on the complex unit circle—thus providing an explicit, time-structured realization of what remains unspecified in Barandes' phase term. Moreover, the rotation can be approximated by a constant angular momentum over the course of a single day.

In this model, all relevant relations enter simultaneously, and the resulting structure aligns fully with the generalized Θ-matrix.

Importantly, the Merge Model developed in our framework also demonstrated a significance level below 1%, further corroborating the consistency and empirical grounding of the approach.

For more clarification, the following table collects notation and meanings (indices are written as α, β rather than i, j to avoid confusion with the imaginary unit i):

Note: Jacob Barandes adopts an uncommon index convention, where $\mathbf{\Gamma}_{ij}(t)$ denotes a transition from state j to state i. We follow the standard notation, in which $\mathbf{\Gamma}_{\alpha\beta}(t)$ refers to a transition from present state α to future state β at time t.

We propose that the review of probabilities in our study, combined with the structural insights from zodiac relations, could serve as a prototype ontology for Barandes' matrix Θ.

Table 5: Comparison of Notation–Barandes vs. This Work

Symbol	Jacob Barandes	This work	Meaning
$\omega_{\alpha\beta}(t)$	Unitary phase factor $\exp(i\theta_{\alpha\beta}(t))$	Time phase	Encodes temporal alignment or structure
$r_{\alpha\beta}(t)$	Transition probability	Importance	Weight of the transition; squared modulus of amplitude
$\Theta_{\alpha\beta}(t)$	$r_{\alpha\beta}(t) \cdot \omega_{\alpha\beta}(t)$	Time number	Encodes time-weighted structural modulation

In this view:

- **The relevant relations** have been assigned numerical values that consistently lead to a frequency of successes, which cannot be explained by chance.
- **Number assignments** establish a relational framework comparable to the probability distributions found in quantum systems.
- **Temporal structures** (e.g., astronomical alignments) function as non-trivial constraints, shaping outcomes in a way reminiscent of hidden variables.

Building on these findings, future research could explore the following:

- **Generalizing the Train Station Model (TSM) beyond roulette**. Applying it to other domains where stochastic processes interact with structured time-dependent frameworks.
- **Formalizing a mathematical bridge between TSM and Barandes' Θ-matrix**. Defining an ontology that unifies stochastic structures with event-based time models.
- **Investigating whether chaotic dynamics in physical systems exhibit similar phase dependencies**. Reinforcing the hypothesis that certain probabilistic outcomes are shaped by underlying structural constraints.
- **Meaning of complex numbers.** The reviewed probabilities, which we refer to as *Importance*, together with phases, naturally form complex numbers. This leads to the understanding that *complex numbers are the natural representatives of a real-time structure*. Consequently, we define them as *Numbers of Time* or *Time Numbers*. For further details, mathematical examples and arguments see [1].
- **We have demonstrated a macroscopic quantum-like model involving hidden variables.** Our results suggest that time is not merely a passive parameter but an active structural component that guides stochastic processes in a meaningful way. In this light, the relation between time, probability and structure may need to be reconsidered—not as a theoretical construct, but as an observable and testable phenomenon with deep implications for physics and beyond.

Thus, in addition to the reasoning above, we can now understand why complex numbers are fundamental in quantum mechanics. The wave function emerges as the ontological representation of a real, existing time structure which cannot be fully captured by classical parameters such as position, momentum or Einsteinian time.

This perspective offers a new interpretation of the mathematical formalism of quantum mechanics, suggesting that complex numbers encode intrinsic temporal structures rather than functioning solely as computational tools. Instead, they inherently represent temporal structures, indicating a deep ontological role in quantum stochastic processes.

We will discuss these topics in a follow-up article.

10.2 Implications and Theoretical Consequences

Finally, our results also call into question the conventional separation between macroscopic and microscopic domains. If the observed regularities in the data are manifestations of structured time, then such structures may permeate all scales—rendering the distinction between "quantum" and "classical" more a matter of observational resolution than of fundamental principle. Importantly, this distinction has never been a true separation between macrocosm and microcosm, but rather between **Einstein events**—observations within a mechanical world—and **chaotic events**, which introduce structural complexity that resists purely dynamical continuation.

This also challenges the very premise of a “Theory of Everything” (ToE) in which general relativity is expected to be quantized. For if structured time governs both chaotic and regular phenomena alike, then the large-scale universe already exhibits analogues of quantum behavior—**not through miniaturization, but through irreducible complexity** emerging when smoothing procedures are suspended and the full dynamics is taken seriously.

10.3 Time Quality and the Relativity of Eventual Time

In Chapter 7, we defined a search algorithm for identifying an essential relation r at an arbitrary Einsteinian time t. If such a relation r was found on day d and matched the time of the first roulette coup on that day, we assigned a strategy S_d using roulette-specific number assignments [11] corresponding to r. Across 4,076 days, such a strategy was identified. Fictional betting based on S_d yielded cumulative wins that could occur by chance only with a probability of $p = 0.00001$ (see Sec. 11.2). Thus, there is evidence of a correlation between astronomical time structures and roulette outcomes.

In general, let an experiment E have outcomes in a given basic set G. For each essential relation r_n, let there be a preassigned subset $Qn \subseteq G$, n=1,…,N. For an arbitrary Einsteinian time *t*, the search algorithm (see Chapter 7) selects an essential relation r_n, which means that we assign the corresponding subset Q_t to time *t*. Furthermore, let a probability measure P describe the outcomes of E. Then a statistical test can be performed with the null hypothesis

H_0: Pr(Outcome at time $t \in Q_t$) = P(Q_t).

If H_0 is rejected, each Q_n, n=1,..,N, is called a *time quality*. The described procedure is called *Time Selection (Zeitwahl).*

Since the choice of time qualities depends on the experiment, they are inherently relational. Hence, whenever statistical significance is achieved, eventual time is relational.

10.4 Example (QM)

Consider interference patterns at a double slit. Let an interference pattern be denoted by I.

1. Truncate the outer tails of I so that 95% of the total mass remains.
2. Normalize the pattern to unit mass.
3. Measure the height a and the width b, and assign the pair (a,b) to I.

By applying Zeitwahl, one performs N distinct experimental runs, each restricted to start times belonging to a specific relation r_n, thereby obtaining the corresponding pattern I_n is represented by its pair (a_n,b_n), n=1,…,N,. Collect these pairs in the set

$$G = \{(a_1,b_1),\ldots,(a_N,b_N)\}.$$

We now formulate the null hypothesis:

H_0: (a_n,b_n), n=1,…,N, are independent samples from the same distribution as (a,b).

The time-choice/quality classification based on globally defined astronomical time structures has so far reached high statistical significance only under the condition of a single, well-controlled trial per day. Empirically, on the order of ~4,000 valid cases are required for a decisive test so a naïve single-site implementation would take decades. To increase statistical power, the experiment should be parallelized across multiple, ideally independent, locations (see [2], Sec. 10.5).

Grouping solely by relations appears plausible, given its success in structuring random outcomes in the roulette case. If this approach does not yield sufficient results, other alternatives can be tested. Before listing them, it is useful to outline a laboratory analogue that translates the roulette scheme directly into an optical setup as discussed earlier in [1], Chapter "Adäquanz der Wohltemperierten Zeit in der Quantenmechanik", Section "Experiment 2: Interferenz".

Consider a flat cylindrical mask featuring 37 equally spaced radial slits (2π/37), labeled as on a roulette wheel, inserted into a coherent photon beam. Each photon passes through one of the slits, and the emerging intensity distribution can be recorded on a detection screen in the Fourier plane. The 37 slits correspond to the set of roulette numbers {0,…,36}. Just as roulette numbers were grouped into relational subsets Q_n, one may predefine angular subsets of slits and group photon detection events accordingly. By performing daily single-run experiments and assigning the results to the relevant subsets, one obtains a statistical dataset directly analogous to the roulette trials.

This design shows that relational grouping and phase grouping are not mutually exclusive: the slit structure encodes angular phases, and the roulette-like number assignment provides the relational coding the relational coding. Both perspectives are valid and can be tested empirically.

The following list summarizes general options for temporal coding. They are of broad scope, yet in point 4 we explicitly connect them to quantum mechanics. Our claim is that the elementary techniques of astrology constitute a more comprehensive form of the event-based logic underlying quantum mechanics.

On this basis, the following alternative groupings can be enumerated:

1. **By relations**: Subsets Q_n as defined in the roulette analogue.
2. **By houses**: Record which house index n determined the essential relation and group results by n.
3. **By ecliptic positions**: Group directly by angular segments along the ecliptic.
4. **By combination of relations and houses**

   The idea here may be that the (constant) initial phase, defined by the initial wavefunction $\psi(\boldsymbol{x_0}, 0)$ can be separated into ϕ_1 and ϕ_2. An elementary wave then takes the form $\psi(\boldsymbol{x}, t) = Ae^{-\frac{i}{\hbar}(E(t-t_0)+\phi_1)} \cdot e^{\frac{i}{\hbar}(\boldsymbol{p}(\boldsymbol{x}-\boldsymbol{x}_0)+\phi_2)}$.

Following the above, we identify the house position with ϕ_1 and the ecliptic position with ϕ_2. In standard quantum mechanics, the position representation to both space and time. For a simple plane wave, however, one quickly sees that space remains essentially static, while the temporal factor provides the ongoing rotation of the phase. In this sense, the position representation is dominated by time: the pulse of time drives the entire structure.

Our splitting into ϕ_1 and ϕ_2 makes this distinction explicit. The ecliptic term corresponds to a slowly varying, global background, whereas the house term carries the rapidly changing, local modulation—the pulse that triggers events within the broader structure. This mirrors the structure of the position representation in quantum mechanics, where time acts as the dominant pulse while space remains comparatively static.

5. **By radial sectors**: The 37-sector optical experiment provides a direct laboratory analogue mentioned above to roulette-based number assignments.

Among these, the house-based grouping appears most plausible as a universal scheme. However, the relational grouping has already proven effective, and the sector-based experiment shows that both views can be unified under one operational design.

These alternatives are not merely different ways of coding the same background structure; they may well differ in their effectiveness. Some schemes—such as the house-based grouping—are likely to capture the underlying order more directly, while others (e.g. ecliptic segmentation) may yield weaker or noisier signals. The empirical task is therefore twofold: first, to test which grouping delivers the strongest and most reproducible deviations from the null hypothesis; and second, to understand whether these differences point to deeper structural hierarchies within the global temporal framework. Thus, the alternatives should be regarded not as redundant, but as complementary probes with potentially unequal success rates.

In this way, the classification of time qualities evolves into a research program: testing which codings stabilize under empirical scrutiny, and whether their relative success reveals structural levels within the ontology of time — even within quantum mechanics itself.

Everyday Anomalies – Towards a Research Program

Rather than reconstructing full interference patterns, classifying atomic spectra in detail, or collecting exhaustive decay data, it may suffice to identify unusual events. The core idea is simple: through time selection, outcomes are classified only by whether they fall within the expected statistical frame or not. Experiments can then be performed at arbitrary times (as a reference) and at times determined by the selection rule. If unexpected events occur systematically more often under time selection, this would indicate that such "outliers" are not statistical artefacts but signatures of temporal structure.

Part III

Supplementary Studies

Quantum Phases, Lorentz Invariance, and Cyclic Cosmology

Dr. Peter Paul Lehleiter

Independent Researcher and Publisher (self-funded)
Morellstraße 19
86159 Augsburg
Germany

Corresponding author:
Dr. Peter Paul Lehleiter
https://orcid.org/0009-0005-1997-8564
E-mail: Peter.Paul.Lehleiter@web.de

Contents

Abstract

Presented here is a collection of interpretations that build on the first article and take into account the second. The focus falls into two main directions: first, quantum mechanics is reinterpreted as the manifestation of cyclic temporal currents, with relativity theory appearing as its limiting case; second, these ideas are extended to cosmology. In established theory, black holes exhibit characteristic oscillations observable as gravitational waves in the aftermath of mergers (the so-called quasi-normal modes). In our framework, we extend this perspective and regard black holes more generally as oscillatory systems stabilized—and ultimately eroded—by temporal energy. Taken together, the results suggest that the universe is not destined for thermodynamic death but instead follows a cyclical path in which apparent disorder is reabsorbed into coherence and collapse is followed by renewal.

Keywords Time structure · Eventual time · Quantum mechanics · Relativity · Cosmology · Dark matter · Dark energy · Coherence

References

[1] P. P. Lehleiter, *Wie Zeit sich in Ereignissen manifestiert: Ansätze einer abstrakten Chronomorphie*. Augsburg, Germany: Chronomorph Verlag, 2024.

[2] P. P. Lehleiter, *Events and Time – in Events, Time, and Quantum Mechanics Supplementary Material*. OSF Project, 2025. Available: https://osf.io/7pgne (see *UsingOSFProject.pdf* in the root folder for project structure and file descriptions).

[3] D. Hestenes, "Quantum mechanics of the electron particle-clock," arXiv:1910.10478, 2019.

[4] D. Hestenes, "Gyromagnetics of the electron clock," *IEEE Access*, 2024.

[5] D. Hestenes, "The zitterbewegung interpretation of quantum mechanics," *Found. Phys.*, vol. 20, pp. 1213–1232, 1990.

[6] S. Carroll, *Space-time and Geometry: An Introduction to General Relativity*. San Francisco, CA, USA: Addison-Wesley, 2004.

[7] I. Prigogine, *From Being to Becoming: Time and Complexity in the Physical Sciences*. San Francisco, CA, USA: W. H. Freeman, 1980.

[8] D. Bohm, "A suggested interpretation of the quantum theory in terms of 'hidden' variables I," Phys. Rev., vol. 85, pp. 166–179, 1952.

[9] Y. Kurihara, "Symplectic structure for general relativity and Einstein–Brillouin–Keller quantization," *Class. Quantum Grav.*, vol. 37, no. 23, p. 235003, 2020.

[10] E. J. Copeland, M. Sami, and S. Tsujikawa, "Dynamics of dark energy," *Int. J. Mod. Phys. D*, vol. 15, pp. 1753–1936, 2006.

[11] G. Ecker, *Teilchen, Felder, Quanten: Von der Quantenmechanik zum Standardmodell der Teilchenphysik*. Springer Spektrum, 2017. doi: 10.1007/978-3-662-54550-8

[12] R. M. Wald, *General Relativity*. Chicago, IL, USA: Univ. Chicago Press, 1984.

[13] Mach, E., *The Science of Mechanics: A Critical and Historical Account of Its Development*. Leipzig: F. A. Brockhaus, 1883.

[14] Griffiths, D. J. (2017). *Introduction to Electrodynamics* (4th ed.). Cambridge University Press.

[15] S. Weinberg, *Cosmology*. Oxford, U.K.: Oxford University Press, 2008.

[16] S. W. Hawking, "Particle creation by black holes," Commun. Math. Phys., vol. 43, pp. 199–220, 1975.

[17] E. Berti, V. Cardoso, and A. O. Starinets, "Quasinormal modes of black holes and black branes," *Class. Quantum Grav.*, vol. 26, p. 163001, 2009.

[18] K. S. Thorne, "Gravitational radiation," in *300 Years of Gravitation*, S. W. Hawking and W. Israel, Eds. Cambridge, U.K.: Cambridge University Press, 1987.

[19] M. Maggiore, *Gravitational Waves: Theory and Experiments*. Oxford, U.K.: Oxford University Press, 2008.

Preface to Part III (Purpose & Scope)

This essay does not aim to deliver a finished cosmology. Its purpose is to widen the conceptual field: to treat time not as a neutral parameter but as a real structuring agent with testable consequences. If physics appears to stagnate, it is because we have optimized technique while narrowing ontology. Part III is a proposal to re-open that space—by reading quantum mechanics, relativity, and cosmology through the lens of time.

What this is / What this is not

What this is. (i) A heuristic framework to reinterpret known structures via time-energy and event-oriented dynamics; (ii) a set of concrete test hooks (Parts I–II); (iii) an invitation to explore, not to settle.

What this is not. (i) Not a Theory of Everything; (ii) not a rejection of established formalisms, but a re-reading; (iii) not a claim beyond the evidence—speculative where marked.

Why now?

Time has been treated as bookkeeping; phenomena we cannot place are renamed "information." We suggest that this is a symptom of conceptual narrowing, not of explanatory success. Re-centering time restores a route to new hypotheses, new experiments, and a different cosmological narrative.

1 Introduction

1.1 General Remarks

This work proposes a heuristic framework of thought in which time is treated not as an external parameter but as a structured, event-bound quantity. Rather than offering a finished physical model, the aim is to demonstrate that such a re-interpretation of time provides a fruitful agenda for further research in quantum mechanics, relativity, and cosmology.

We distinguish three complementary layers: **Einsteinian Time** as a neutral mechanical scale, **Eventual Time** as the process-linked temporal flow [1,2], and **Potential Time** (introduced here) as the background reservoir from which events are drawn. Within this conceptual setting, quantum mechanics is reinterpreted as the manifestation of circular time currents, while relativity theory appears as its limiting case when the effective action quantum becomes large.

Extending this approach to cosmology, black holes are understood as oscillatory systems stabilized and ultimately eroded by temporal energy. Dark matter and dark energy then emerge as complementary expressions of the same temporal substrate: stored (static) and expanding (kinetic) forms of temporal energy. From this perspective, the universe follows a cyclical–evolutionary path in which apparent disorder is reabsorbed into coherence and large-scale collapse is followed by renewal.

The present volume does not claim to deliver a closed theory but presents a set of related interpretations. These build on the first article and take into account the second, converging into two main directions: the kinship of quantum mechanics and relativity, and—on a larger scale—the cosmological implications.

1.2 Connections to Previous Parts and Basic Discussion

In order to place the present cosmological considerations in context, three points from the preceding work should be recalled.

- **Evidence.** The second article provided empirical indications that temporal structures are not merely philosophical constructs but have observable effects. By applying time selection to randomness experiments, systematic deviations from pure chance were found [1,2]. This suggests that temporal relations may enter physical processes in a measurable way.
- **Possible Experiments.** The first article outlined concrete experimental approaches to test this claim: interference patterns probing phase coherence, spectral line shapes probing energy–phase correlations, and decay statistics probing deviations from Poissonian models. These proposals remain programmatic but illustrate that the temporal interpretation can be subjected to empirical scrutiny.
- **The Bohmian Quantum Potential.** A key conceptual link is Bohm's quantum potential (QP). In standard treatments it appears as a correction term at the microscopic level. In our framework, however, it is not a small-scale anomaly but a universal feature: the QP arises wherever structure and richness of correlations occur, including chaotic processes. This view is motivated by Chapter 3, where we argue that space-time in special relativity can be understood as a limiting case of a wave function. Since the QP is derived from the wave function and the Schrödinger equation, it represents a statistical or averaged quantity. This averaging aspect opens the way to a first definition of *temporal energy*: at the quantum scale it appears as averaged mechanical energy, while the underlying phases are more fundamental. The equation for the quantum potential is coupled with the continuity equation,

so the modulus of the wave function is not independent of the temporal structure expressed by the phase. More generally, the central role of the phase in superposition, interference, and entanglement strongly suggests that the quantum potential encodes temporal energy. In this sense, time-energy can be given a physical grounding.

- **Relation to Chapter 3.** As argued in Chapter 3, space-time itself may be understood as the limiting case of a wave function. From this perspective, Bohm's potential is not confined to the microscopic regime but expresses a general structural principle of physics. At small scales its temporal character is visible only in statistical averages, while at larger scales this restriction falls away. The empirical studies on structured randomness (e.g., roulette experiments) may therefore be regarded as pilot studies for a physical concept of temporal energy.

1.3 Temporal Energy and the Dark Sector

From our perspective, space–time is indivisible, and energy must be classified into spatial and temporal forms. Spatial energy manifests in matter and radiation, i.e. in localized mechanical entities. Temporal energy, by contrast, acts globally and does not reduce to mechanical components. In this light, the enigmatic phenomena of dark matter and dark energy may be viewed as temporal manifestations rather than spatial ones. The detailed argument for this reinterpretation—including its cosmological implications—is developed in Sec. 4.9.

2 Spin as an Expression of a Cyclic Temporal Quality

Hestenes' interpretation (see [3,4,5], [2] in Section 6.3) suggests that the *zitterbewegung* defines an *internal time* for the electron. We now go one step further in our interpretation of spin.

Spin stands out as a defining feature of QM because the characteristic "two-turn" behavior of spin-½ systems does not represent a physical rotation in space but rather a kind of *temporal bookkeeping*: an intrinsic, cyclic internal time structure that expresses a single quality alternating between two opposing valuations.

The system switches between these opposing states, and only after two successive transitions—i.e., completion of the full cycle—is the original valuation restored. Spatial depictions of spin serve as formal encodings; they do not reflect the true ontological content.

This alternation defines a minimal internal time cycle: each change of valuation marks a temporal event, and the cycle restores the prior state—like alternating ledger entries. Externally, this appears as the spinor phase: a rhythm of qualitative opposites that closes only after two steps. The alternation serves as a temporal marker, reflected in the double-valuedness of spin-½. Spin is thus closer to a qualitative rhythm than to a geometric rotation, strongly suggesting that QM is intrinsically structured by internal temporal dynamics—beyond Einsteinian time parameters and grounded in event-driven time.

3 Lorentz-Invariant Phases and the Temporal Core of QM

For the basic notions of Lorentz invariance and relativistic invariants, we follow standard expositions [6]. As outlined in Section 9.2, the method presented in [1,2] introduces a temporally ordered structure based on ecliptic positions. This structure is subsequently integrated into the quantum-mechanical framework (see Sections 10.1–10.5). Since this approach involves cyclic phases φ_1 and φ_2 at Einsteinian Time t_1 and t_2, specifically within the interval $[0,2\pi]$ we may

$$\phi = \frac{\Delta S}{\tilde{\hbar}} \tag{11.1}$$

and interpret $\tilde{\hbar}$ as a *mesoscopic or macroscopic action quantum* that governs coarse-grained phase dynamics in this context.

This leads us to a more general hypothesis: according to the principle of well-tempering within the holoperspective worldview, different levels of physical structure may require different effective values of $\tilde{\hbar}$. At the quantum level, of course, we recover the standard identity $\tilde{\hbar} = \hbar$,

but on higher (mesoscopic or macroscopic) levels, alternative phase constants may be more adequate descriptors of structured temporal dynamics.

We write the corresponding elementary wave function as

$$\psi_k = e^{-\frac{i}{\tilde{\hbar}}(Et_k - \boldsymbol{p}\cdot\boldsymbol{r}_k)}, k = 1{,}2.$$

After Definition 2.3 we compare evaluated measures like ψ_k by using (generalized) Born's Rule:

$$\langle\psi_1|\psi_2\rangle = e^{-\frac{i}{\tilde{\hbar}}(\mathrm{E}\Delta t\, -\, \boldsymbol{p}\cdot\Delta\boldsymbol{r})} = e^{-\frac{i}{\tilde{\hbar}}\varphi} \tag{11.2}$$

Applying the first-order Taylor expansion $e^{-y} \approx 1 - y$ (for $|y| \ll 1$) to equation (11.2) with $y = \frac{i}{\tilde{\hbar}}\varphi$ gives

$\langle\psi_1|\psi_2\rangle \approx 1 - \frac{i}{\tilde{\hbar}}(\mathrm{E}\Delta t \; - \; \boldsymbol{p}\cdot\Delta\boldsymbol{r}).$

Since operating in the exponent with a small argument effectively yields a unit shift, we omit this constant term as physically irrelevant for phase comparisons. Furthermore, we eliminate the overall scale factor, which merely fixes the units and does not affect the structural form of the phase-dependent term. Having thus reduced the quantum expression to its classical limit, we no longer speak of phases but of the corresponding relativistic invariant. We now consider a relativistically relevant scenario, in which the particle follows a classical trajectory, such that $\Delta\boldsymbol{r} = \boldsymbol{v}\Delta t$. The Lorentz-invariant action scalar φ (with bold symbols denoting vectors) is then given by using the Lorentz factor γ:

$$\varphi = E\Delta t - \boldsymbol{p}\cdot(\boldsymbol{v}\Delta t) = \gamma mc^2\Delta t - \gamma mv^2\Delta t$$

$$= \gamma m(c^2 - v^2)\Delta t = \gamma mc^2\left(1 - \frac{v^2}{c^2}\right)\Delta t = \gamma m\frac{c^2}{\gamma^2}\Delta t = mc^2\frac{\Delta t}{\gamma} = mc^2\Delta\tau,$$

where $\Delta\tau = \Delta t/\gamma$ is the proper time. It follows that

$$|\varphi| = mc^2\Delta\tau.$$

For $\Delta\tau \ll \frac{\tilde{\hbar}}{mc^2}$, the scaled phase satisfies

$$\frac{|\varphi|}{\tilde{\hbar}} = \left| mc^2\frac{\Delta\tau}{\tilde{\hbar}} \right| \ll 1.$$

In four-vector notation:

$$\varphi = \boldsymbol{p}_\mu \, \Delta\boldsymbol{x}^\mu = \mathrm{E}\Delta t - \boldsymbol{p}\cdot\Delta\boldsymbol{r}$$

which is manifestly Lorentz-invariant under transformations of SR, provided that $\Delta\boldsymbol{x}^\mu$ refers to the increment along the particle's worldline (so that $\Delta\boldsymbol{r} = \boldsymbol{v}\Delta t$).

Remarks

- If the chronomorphic perspective [1,2] is globally confirmed, both QM and special SR can be understood as emergent theories—arising from a deeper temporally relational model based on rotational dynamics.

SR appears as an effective limiting case, since along the particle's worldline $|\varphi|$ $= mc^2 \Delta\tau$ while $\tilde{\hbar}$ is macroscopically large.

In this view, the formal similarity between the Lorentz-invariant phase φ in SR and the complex phase Φ in QM is no coincidence; both arise from the same underlying rotational time structure. As the degree of mechanicality increases, i.e., as $\tilde{\hbar}\rightarrow\infty$, the event-based character of the theory diminishes, and QM smoothly transitions into SR.

- In this limit, the cyclic evaluated time measures converge to a flat Minkowski space–time.
- The change in the roles of space and time at the event horizon of a black hole is analogous to this transition: just as crossing the horizon interchanges spatial and temporal coordinates, the Schrödinger equation—through the factor i in its time evolution—links the drift of the system to its cyclic phase structure, coupling energetic and temporal aspects within a single complex dynamic. In this process, the drift becomes linear and the cycle itself flattens, reducing time to a single axis in the SR sense. In both cases, what appears as a coordinate transformation reflects a deeper structural shift in the underlying temporal geometry.
- This parallel points to a deeper, universal principle: a fundamental harmony condition linking temporal and energetic structure. Within this framework, quantization is not the starting point but an emergent consequence of this harmony—a manifestation of the Principle of Well-Tempering—present across all scales. Even in the special-relativistic limit, this underlying relation persists, albeit in a reduced form that no longer exhibits overt quantum interference.

From this perspective, QM is already encoded within SR through the same underlying angular dynamics, though it remains hidden in the absence of non-mechanical phenomena causing significant quantum effects. This suggests that the classical pursuit of a unified Theory of Everything (ToE) is misguided: QM and SR are fundamentally incompatible, because both are complementary projections of the same underlying temporal structure. More broadly, physics neither needs—nor is able—to subsume all domains into a single formal framework. Distinct theories can coexist, each valid within its own domain, without enforcing reductive unification. Even phenomena such as perception or primitive forms of consciousness, which may leave traces in quantum dynamics, belong to separate explanatory layers. The insistence on merging them all into a single edifice risks turning a provisional construct into an obstacle to deeper insight. Especially, it is misguided to accept only those theories that contain, for example, QED as a subset; there is likewise no necessity to extend the de Broglie–Madelung–Bohm theory to incorporate aspects of QED. Moreover, the Bohmian quantum potential [8] may prove useful in gaining further insight into the new perspective on time presented here, as it could represent the energetic effects associated with phase, hence with time—though, in this framework, phase would not be merely postulated as an initial stochastic condition, but seen as arising from global structures that act as coherence constraints. Since, ultimately, nothing in physics is expressed in terms other than energy and/or matter, this link remains a promising avenue for further exploration.

As long as theoretical development is constrained within the patchwork corset of QED, physics will remain limited to familiar steps—never venturing into new choreographies that could creatively advance the discipline.

The concept of evaluated time measures may initially appear abstract. However, this is no more problematic than accepting that a single Coulomb charge instantaneously defines an electric field throughout all space in the static case. If such nonlocal field concepts are fundamental to physics, evaluated time measures deserve equal consideration as legitimate physical constructs.

As already discussed in Section 5.2, the EBK framework captures the inherently circulatory character of quantum wave functions, reducing quantization to phase conditions on closed trajectories. Notably, as Kurihara has recently shown [9], the same symplectic framework can be formulated for general relativistic systems, where EBK-type quantization rules apply to geodesic motion. This parallel suggests that the cyclic temporal structure underlying semiclassical quantization is not confined to quantum mechanics but manifests equally in relativistic dynamics. From our standpoint, this parallelism does not open the door to a "grand unification" of QM and SR in the usual sense. Rather, it emphasizes their shared root: both theories emerge from a deeper cyclic temporal structure. In this reading, EBK quantization appears as a universal expression of circular time—bridging microscopic phases and macroscopic trajectories. Thus, quantum mechanics and relativity are not to be unified but to be re-read as complementary projections of time itself.

Dark Energy, SR and QM

In the temporal interpretation, dark energy can be understood as an expression of time itself. Minkowski's space–time continuum may be viewed as a degenerated wave function, and interpreting it as a temporal flow suggests that cosmic expansion is grounded in a time-like process. Since no spatial manifestations such as mass, radiation, or particle byproducts are observed, the only remaining component in the relativistic mass–space-time framework is time. Dark energy therefore be understood as a temporal form of energy—Potential Time—which not only gives it a natural structure consistent with relativity and quantum mechanics but also may even shed light on cosmological origins. For reviews of conventional dark energy models, see [10].

4 Cosmology – A New Approach

The expansion of the universe is often portrayed as a tragedy: an ever-thinning cosmos heading toward thermal death. We suggest the opposite. Expansion is not a flaw in the parameters, nor a cosmic nuisance, but the very trajectory toward coherence. Dilution of chaotic events and the decay of structures are not signs of loss; they are the preconditions for the universe to inscribe its final meaning. What is called the Big Freeze may thus be understood as the testament of the universe, in which its essential order is preserved for a new beginning. The perspective changes once one recognizes that the endpoint is not emptiness but coherence.

Atoms consist predominantly of empty space. The atomic radius is about 0.53×10^{-10} m, whereas the nuclear radius is of the order of 10^{-15}m [11]. Thus, the atom is larger than its nucleus by a factor of approximately 10^5. Einstein himself provided evidence for this shift. Though celebrated as the master mechanic of physics, his general theory of relativity accomplished the opposite: it removed Newtonian force from the stage and replaced it with pure geometry [12]. Gravitation as a force was not explained but rendered unnecessary by the recognition of underlying structure. In this sense, order precedes mechanics, and the supposed foundations of physics reveal themselves as consequences of something deeper.

The same lesson appears in other domains. Inertia, for example, is often explained as the effect of all masses in the universe, a thought related to Mach's principle [13]. Yet gravity is defined mechanically, through force and acceleration, while inertia is defined as resistance to acceleration. Mechanics here explains mass by force and force by mass—a circular reasoning that cannot close within mechanics itself. What remains is that inertia reflects the global order of matter in the universe. Electrostatics makes the point even clearer: a neutral region has no field, but the instant a charge exists, all space is instantly structured by its field. Place two opposite charges, and their superposed fields pervade all space without delay. This is not dynamical propagation but the order of space itself. Coulomb's force is only the mechanical derivative of this prior structuring [14]. Once again, mechanics is not fundamental but rests on premises that are radically non-mechanical: order precedes mechanics.

4.1 Introduction: Basic Assumptions

By *event* we mean not merely a state change or the manifestation of a mechanical property, but an occurrence that constitutes a structural break or the sudden emergence of a random influence. Events in this sense are linked to irreversibility: once they happen, the system cannot return to its exact prior state.

Our cosmological model rests on the following assumptions:

- **Time appears in three complementary versions**
  - **Einsteinian Time**

    This is the time shown by clocks. It provides an external scale, neutral to any process. The fact that completely different processes can be measured with the same clock reveals its mechanical, indifferent character. Since mechanical dynamics consist only of reversible state changes, they are eventless in our sense. Einsteinian time thus represents the limiting case of such eventless, purely mechanical processes—an abstraction that isolates the reversible side of temporal phenomena.

    In short: Einsteinian Time is merely the abstract limiting case of eventless mechanics.
  - **Eventual Time**

    In contrast, Eventual Time is inseparable from the process itself. It manifests in the unfolding of the process—especially when that process is rich, stochastic, or unpredictable. In this sense, the process itself becomes its own clock. Eventual Time makes itself tangible in the occurrence of events.

In short: Eventual Time manifests itself in the unfolding of irreducible events.

- **Potential Time**

  We further assume that Eventual Time is grounded in an underlying dimension, Potential Time. It functions as a reservoir of contents that become actualized in events: Eventual Time manifests processes by drawing from Potential Time into the present.

  In short: Potential Time is the latent dimension from which events draw their actuality.

- **Space–time follows two holistic principles:**

  In our investigations ([1]; Sec. 9.2 in [2]) strong evidence has been found that global temporal structures manifest within local events. This suggests that part of what appears as mere randomness must instead be understood as an order reflecting larger relations. From this observation, the following two principles are derived:

  - *Holoperspectivism* – the universe is holistic in the sense that global relations become manifest within local events.
  - *Well-Tempering* – even from chaotic processes, coherence emerges anew; this principle of order prevents the universe from disintegrating into mere chaos.

- **Principle of Coherence**

  Building on holoperspectivism and well-tempering, our investigations suggest that the observed patterns are not confined to isolated events but point to a pervasive principle: the universe as a whole tends toward maximal coherence. What appears as emerging randomness is thus continually reorganized into new forms of order. Energy—and therefore mass—sets barriers against this tendency, creating and stabilizing structures. Yet these barriers are not permanent: as structures decay, coherence reasserts itself and increases on a higher level.

- **Time is real**
- Behind the space–time coordinates lies a process of real time. Accordingly, Minkowski's metric [6] is not merely a mathematical device but should be taken seriously as the representation of an ontological object.
  Throughout these articles, my guiding paradigm has been to read formulas not merely as computational devices but as structured expressions that carry meaning. Rather than reducing them to algorithms, my intention is to read them in a way that allows their ontological implications to become visible. In this sense, Minkowski's metric can be read as an ontological object.
  The energetic aspect of time becomes tangible in quantum mechanics: on average, it is expressed through Bohm's quantum potential, which can be derived directly from the Schrödinger equation and thus provides an immediate pointer to this hidden energy. As shown in Chapter 3, Minkowski's metric can be regarded as the flat limiting case of an elementary wave; traces of such a "time-energy" can therefore also be discerned there. Here and below, "time-energy" is meant in an applied-GR sense as part of the GR calculus, not as a new substance.

Illustration of the time concepts

A film unfolds according to a script. The script is a notebook of numbered pages that could have been filled with any content—recipes, diary entries, or observations.

Imagine a whole library of such notebooks, written in an unknown language, so that only their broad structure is accessible.

- The page numbers correspond to Einsteinian Time: external, eventless, meaningless, with no reference to content.
- The plot of the film corresponds to Eventual Time: the story in its scenes, twists, and dynamics—the decisive turns that shape the drama.
- The library of notebooks corresponds to Potential Time: a diversity of contents, symbolizing possibilities not directly accessible.

The most faithful process time is the process itself—the process as its own clock, which we call a *Chronomorph* [1]. Yet such a notion, while conceptually illuminating, offers little practical use. Any operational notion of process time must therefore be coarsened: the process must be reduced to a substructure that serves as a temporal marker.

In dissipative systems, substructures of process time are given by successive order regimes. When the energy supply increases, existing configurations reach their limits, and the transport of energy must be reorganized. The system then enters a new regime, characterized by patterns suited for higher throughput [7]. Conversely, when the energy input decreases, earlier regimes may reappear. This does not undo the microscopic arrow of time: dissipated energy cannot be restored, and the past does not return. What recurs are structural qualities—macroscopic patterns that time itself preserves.

Such regime changes provide a coarse but consistent measure of Eventual Time, perhaps the coarsest conceivable one. They also suggest a deeper implication: if earlier regimes can re-emerge, the first emergence of new regimes must follow the same principle. Their structures, in some sense, pre-exist—"stored" in time—so that they can either recur or arise for the first time within the process. As Prigogine emphasized, far from equilibrium matter may spontaneously form ordered regimes, maintained by the very flow of energy through the system. Their origin cannot be predicted from the previous state alone; fluctuations, amplified by time's irreversibility, select the outcome [7, Chap. 3]. Dissipative structures thus point to a form of "time storage," where temporal order provides the background from which structures emerge and to which they can return.

Remarks:

- The recurrence of structural qualities is not limited to physics. In broader contexts it suggests that temporal patterns may accelerate processes such as biological evolution, where established qualities can reappear and be integrated at higher levels of organization.
- We do not insist on a single vocabulary. Readers may adopt other models or employ different terms, or even prefer an alternative ontological framing. What matters is not the label, but the recognition that beyond the standard physical formalism, additional processual entities are required.
- The empirical dimension of this program has already yielded promising statistical results, as discussed in Part II. These findings can be further refined by coordinated experimental designs that aggregate data across different types of trials, including in quantum mechanics. While a detailed technical plan will be developed separately, it is clear that systematic replication and diversification of approaches provide the option to strengthen and expand the evidence base.
- Historically, the privileging of a purely mechanical notion of time has shaped scientific practice and culture in ways that tend to marginalize temporal or processual notions. This marginalization has consequences beyond vocabulary: it influences what questions are asked, which phenomena are

deemed worthy of study, and how technical power is deployed. Our project proposes a reconsideration of that legacy by restoring the temporal dimension to a central role in theory and experiment.

Taken together, these assumptions establish a framework for reconsidering cosmology from first principles. Time is no longer treated as a neutral background parameter but as a real structuring agent that shapes processes, coherence, and cosmic evolution. The complementary notions of Einsteinian, Eventual, and Potential Time, combined with the principles of holoperspectivism and well-tempering, offer a unified perspective: local events reveal global order, apparent disorder tends back toward coherence, and scenarios of cosmic decay appear as limiting cases. Within this perspective, a cyclic view of the universe emerges naturally—as the consequence of real time and its inherent drive toward coherence.

4.2 Expansion and the Coherence Principle

The assumptions outlined above provide a framework for rethinking time, process, and structure in cosmology. But what does this mean for the evolution of the universe as a whole? To illustrate, let us consider the phenomenon of cosmic expansion.

Coherence is not a local accident but a universal principle. Its manifestations range from the atomic to the cosmological scale. What appears microscopically in Ritz's combination principle—the selective retention of certain frequencies—already reflects the same structural ordering that, on the largest scales, governs cosmic expansion. In our perspective, the distinction is not between micro and macro, but between mechanical evolution without events and event-oriented structures shaped by coherence. Seen this way, Ritz's rule is not a technical restriction but a local trace of a universal ordering principle [2].

The expansion of the universe is not a misalignment of parameters, not a cosmic nuisance. It is the path back to fundamental coherence. This requires the dilution of chaotic events and the dissolution of structures. Already in the so-called Big Bang this direction was inscribed: from the very beginning, the trajectory pointed toward coherence.

Time and meaning are inseparable. Every relation in time can carry meaning—demonstrable in the layering of interference patterns ([1, Sec. xx]). It would therefore be absurd if space-time were to expand without carrying meaning. Rather, in its expansion the universe inscribes its own meaning—like a hologram recorded with a coherent frequency and later made visible again.

In the chaotic phases, as long as the universe is not yet fully expanded, the final meaning of the universe is inscribed.

It is essential to distinguish between local turbulence and coherence on a higher scale. Brownian motion, with its dense superposition of countless collisions, exemplifies the former: the sheer abundance of processes at one place and time erases any larger relation, producing the appearance of pure noise. Coherence does not emerge at this level but only on an overarching scale.

This explains why mesoscopic experiments, where individual events occur sparsely and are separated in time, can reveal correlations with global temporal structures. When too many processes overlap locally, interference among them conceals coherence; but when events are rare enough, the connection between local randomness and cosmically defined rhythms can be traced.

From this perspective, a purely thermodynamic or stochastic account remains insufficient, since both presuppose strict mechanics and exclude the action of time itself. What seems like agitation without meaning is, on the higher scale where density is reduced, the very spacing that allows coherence to become visible. The expansion of the universe magnifies this effect: as structures dissolve and disperse, remaining events become increasingly isolated, until their coherence with the whole stands out.

This coherent energy appears cold because it is spread so widely, yet it persists. By means of a Fourier transformation, this energy is concentrated again into a narrow spatial region. From this concentrated core emerges the next beginning: what later researchers interpret as the Big Bang is in truth the folding back of the final coherence of the old universe.

Thus we understand: the universe is not empty mechanics, not bizarre dead matter. Everything that exists carries meaning within itself—at least in the reductive sense that meaning lies in its very existence. Expansion is not a flaw of parameters but part of the cosmos from the outset. It is the long developmental path that allows structures to arise and decay, only to culminate once again in a new beginning.

4.3 From Quantum Mechanics to Cosmology

Quantum mechanics can, in our approach (Part I), be understood as an expression of Eventual Time. While in the standard interpretation randomness is regarded as an intrinsic property of quantum mechanics—and thus of the universe—we conceive the essence of quantum mechanics as a flow of time itself. This becomes immediately apparent in the Schrödinger equation:

$$i\hbar\frac{\partial\psi}{\partial t} = (-\frac{\hbar^2}{2m}\nabla^2 + V)\psi. \tag{5.1}$$

If rewritten so that the time derivative is isolated, one obtains:

$$\frac{\partial\psi}{\partial t} = \frac{i}{\hbar}\widetilde{\widetilde{H}}\psi,\ \widetilde{\widetilde{H}} = \frac{\hbar^2}{2m}\nabla^2 - V. \tag{5.2}$$

The right-hand side represents a flow quantity: it contains the second spatial derivative with a positive sign, together with a contribution from the potential. Taken together, this yields a complex-valued flow with a clear circular component in the complex plane. We interpret this as a circular time flow.

Thus, in our understanding, quantum mechanics is nothing but a report on the action of Eventual Time ([2, Chap. 3, 5, 6)]. This perspective also allows for a scale generalization. We introduce a reduced action quantum $\tilde{\hbar}$. When it becomes very large, the elementary wave function (see Chapter 3) approaches a description in terms of four-vectors of relativity theory. In this sense, relativity theory appears as a limiting case of quantum mechanics:

$$\tilde{\hbar} \rightarrow \infty \Rightarrow \langle \psi_1 | \psi_2 \rangle \rightarrow 1 - \frac{i}{\tilde{\hbar}}(\mathrm{E}\Delta t - \boldsymbol{p} \cdot \Delta \boldsymbol{r}). \tag{5.3}$$

Already in the early beginnings of quantum mechanics this unity becomes apparent. De Broglie identified in the spectra an intrinsic frequency to be attributed to the electrons: $\nu_0 = \mathrm{m}c^2/h$. Since the spectra had to be independent of the reference frame, he was allowed to apply the Lorentz transformation to this internal oscillation. Stability could be ensured only if the original vibration and the transformed wave remained coherent in the laboratory frame.

4.4 Classical Cosmological Models

In established cosmology, two major end-time scenarios are discussed [15], [6]:

- **Big Freeze**

  This scenario assumes an unlimited expansion of the universe. As the expansion proceeds, matter becomes increasingly diluted, the distances between galaxies grow, and gravitational as well as thermodynamic interactions continue to diminish. At the same time, according to the second law of thermodynamics, entropy increases. Energy differences level out, temperature gradients vanish, and in the limiting case the entire universe approaches absolute zero. Stars burn out, matter decays, and what remains is an almost empty, cold space-time.

- **Heat Death**

> Here too, the increase of entropy is central. In this picture, energy eventually spreads evenly throughout the universe. Once all energy gradients have disappeared, no work can be performed. The universe reaches a state of maximal disorder. "Heat" here does not mean high temperature, but rather a fully leveled thermal distribution. Everything is "used up"—motion, structures, and processes come to a standstill.

What both models have in common is the growth of entropy. Our idea is that increasing entropy goes hand in hand with increasing chaos, visible in the disappearance of structures, leaving behind little more than marginal flickering. However, more and more black holes keep forming. Before merging, they orbit one another and lose energy in the process. Their coalescence generates gravitational waves, thereby releasing further energy.

We now consider whether black holes persist over enormous timescales and, in theory, "evaporate" only through Hawking radiation [16]—or whether their lifetime is in fact significantly shorter.

4.5 Black Holes and Time-energy

Let us first consider what happens inside a black hole. For this, we examine what takes place at the event horizon when an external object encounters it. From a formal–mathematical point of view, we see that time and space exchange their roles [13]. As with the wave function, we take the Minkowski–Einstein space-time ontologically seriously, not merely as a mathematical gimmick, as stated by our basic assumption (see Sec. 4.1). This view is further supported by the fact that, as shown in Chapter 3, one may regard it as the flat limiting case of an elementary wave function. Moreover, our considerations in Part I, Sec. 9.2, starting from the Schrödinger equation, led to the assessment (see Sec. 4.1) that time possesses an energy equivalent. Taken together, it follows that when crossing the event horizon, the energies (the energy equivalent of mass and of time) also exchange their roles. That is, mass is converted into energy, and time-energy into mass. Thus, inside the black hole, a comparatively small amount of mass remains, while it consists predominantly of time-energy.

In this picture, the interior of a black hole can no longer be described in terms of ordinary spatial matter. What remains effective is the temporal component, which carries its own energetic weight. To substantiate this claim, one may recall Einstein's fundamental relation between energy and mass.

4.6 Time-energy as Energy Equivalent

According to Einstein,

$E=mc^2$

every form of energy possesses a mass equivalent and therefore contributes to gravitation. This also applies to time-energy inside a black hole. As long as the internal processes remain disordered and chaotic, their chaos carries the mass equivalent that stabilizes the horizon. The black hole thus persists even in the absence of classical matter.

In this sense, gravitation remains effective when matter energy is transformed into time-energy. Speculatively, this time-energy may be interpreted as a form of "dark matter." For the present argument, however, it is sufficient to note that the gravitational effect follows directly from Einstein's relation.

Readers not inclined to adopt this speculative link to dark matter may nonetheless follow the subsequent argument. In Section 6.4 we will show that even in conventional terms the interior energy can be understood as self-organizing into coherent oscillations, leading to black hole decay.

4.7 The Black Hole as a Resonant Bell, Excited by Eventual Time

As entropy increases, disorder and chaos grow. Yet according to the holoperspectival worldview and the *principle of well-tempering (in analogy to music)*, a global alignment of coherence simultaneously arises. In this picture, *energetic structures—including masses—stand in opposition to coherence, the true ground state of the universe.* With their disappearance, coherence increases.

This coherence sets black holes into oscillation, causing them to lose energy and shrink. The residual matter described in the previous section may be too small to stabilize the black hole. If so, the black hole evaporates through Hawking radiation [12] within only a few dozen or a few hundred million years.

In our framework, the decay timescale refers to the lifetime of a black hole from its very existence, not postponed to the remote epoch of heat death. As argued in Parts I and II, the fundamental tendency of the universe is toward coherence, articulated in two sub-principles: the holoperspectival worldview and the principle of well-tempering. Both become increasingly effective as mechanical order fades and chaos grows. What remains of disintegrated structures is only a chaotic flicker, which is gradually drawn into coherence. It is precisely this growing coherence that drives black holes into oscillation and accelerates their decay.

In the conventional picture, the interior of a black hole is often imagined as a chaotic plasma of energy. Yet just as outside the horizon, chaotic energy is expected to self-organize into coherent structures. These structures render the horizon susceptible to oscillations and external perturbations, eventually leading to decay. In our framework, *time-energy* has a decisive advantage: it is defined through phases and is coherent by nature. Hence black hole decay proceeds more efficiently, without requiring extra assumptions about self-organization.

Building on the considerations in Part I and Part II, the conclusion follows immediately: strict chaos does not exist. Our roulette experiments already demonstrated that seemingly random outcomes align with higher-order structures. The same principle must hold inside a black hole. What appears as a hot, disordered plasma in reality organizes itself into coherent oscillations. These oscillations make the horizon vulnerable to resonances with external fluctuations and ultimately cause the emission of energy.

It would be inconsistent to argue that black holes evaporate faster merely because energy, like mass, gravitates. In Hawking's formula, the equivalence of mass and energy is already assumed. Our reasoning takes a different path: what appears as chaos is never the final state. Disorder represents the decay of an energetically sustained but fundamentally alien order. The true ground state of the universe is coherence, and every apparent chaos ultimately reorganizes itself into coherence. In black holes, this reorientation generates oscillations that destabilize the horizon and lead to energy emission.

The universal constant *h* suggests that coherence is not an occasional feature but a pervasive background. Quantized energy packets can be seen as disturbances or multiples of this underlying coherence. When such structures decay, what remains is not disorder but the re-emergence of the fundamental oscillation of the universe.

This picture does not imply a continuous, monochromatic gravitational-wave tone. Quasi-normal modes are damped (with decay time $\tau \sim Q/\omega$) and are only excited in brief episodes by external perturbations or stochastic temporal forcing; the expected signal is therefore a sparse sequence of short ringdowns with a very low duty cycle $f \ll 1$.

To make the qualitative picture more concrete—without implying persistent oscillations—we record a time-averaged, speculative scaling estimate:

Speculative scaling estimate (time-averaged; for orientation only).

Treat the black hole as a system that is intermittently excited into a dominant quasi-normal mode of angular frequency

$$\omega \approx \frac{c^3}{GM}$$

[16], [17] and quality factor Q [16],

When excited, a fraction $\eta \ll 1$ of the rest energy is stored in this mode and coupled out:

$E_{mode} = \eta Mc^2$ [18].

The instantaneous radiated power during an excited episode is

$P_{inst} \approx \frac{\eta}{Q}\frac{c^5}{G}$ [18].

If such episodes occupy only a small fraction *f* of the time (duty cycle), the time-averaged power is

$\langle P \rangle \approx f\frac{\eta}{Q}\frac{c^5}{G}$ [19],

and thus

$$\dot{M} \approx -\frac{\langle P \rangle}{c^2} \approx f\frac{\eta}{Q}\frac{c^5}{G},$$

and hence the average decay time

$$t_{decay} \approx \frac{G}{c^3}\frac{Q}{f\eta}M_0.$$

With $G/c^3 \approx 2.47\times10^{-36}$ s/kg and $M_0 = 10M_\odot \approx 2\times10^{31}$ kg:

$t_{decay} \approx 4.9\times10^{-5}(Q/(f\eta))$ s.

Values like

$Q \approx 10$, $(f,\eta) = (10^{-10},10^{-6})$, $(10^{-11},10^{-7})$, or $(10^{-12},10^{-8})$

give

$t_{decay} \approx 10^7$ - 10^8 years

while predicting no continuous narrowband emission—only rare, short ringdowns far below current detection thresholds unless strongly perturbed (e.g., during mergers).

Caveat. This is a deliberately speculative toy model: it ignores back-reaction, multi-mode coupling, selection rules, and environment. It illustrates only the plausibility of short cosmological lifetimes under intermittent forcing, not a firm prediction.

4.8 Coherence Catastrophe and Restart

As black holes successively decay, global coherence increases until it saturates. By Fourier concentration, energy then focuses into a narrow spatial region, triggering a sudden restart of the universe.

Yet in the phases of Potential Time lies the still-relevant residual information of the universe, not least through coherence with the Schwarzschild radii of the black holes and through the coherence of the seemingly chaotic process. This information is carried into the reset as a structuring principle of the energy that drives the new beginning. In this way, the universe appears not only as cyclic but also as evolutionary.

4.9 Meaning of Dark Matter and Dark Energy

Hints toward the existence of time-energy are not restricted to quantum mechanics [1,2]. They also arise from the puzzling proportions of conventional cosmology, expressed in the concepts of dark matter and dark energy. What is striking is that these components cannot be assigned to ordinary spatial entities: they exhibit no spatial properties such as radiation or particle-like behavior, which are hallmarks of space-related phenomena. Instead, dark matter manifests only through gravitation, and dark energy only through the metric expansion of space–time.

Since space–time is indivisible and time forms its integral part, the natural conclusion is that both phenomena belong to the temporal aspect of reality. To avoid this step would be to concede that general relativity is incomplete. Yet physics must in any case be extended—so why not take the decisive step now and interpret dark matter and dark energy as manifestations of time itself?

Such an interpretation requires an expanded concept of time and spares us ad hoc assumptions. Einstein himself resolved contradictions in physics by completing mechanics through relativity. Yet the concept of time—and with it the concept of the event—remained largely unexamined. Our approach therefore develops a more general notion of time, one that is intrinsically event-oriented. In this sense, a step parallel to Einstein's achievement in mechanics may now be required in order to advance the foundations of physics.

The same challenge appears in quantum mechanics: conventional descriptions often dissolve into pure mathematics, prompting the claim that information is the fundamental substance of the universe. But information plays a role similar to that once attributed to the ether: a provisional concept covering gaps while lacking true explanatory power. History has shown that the ether was a dead end that cost valuable time. It would not be surprising if "information" were also only a stopgap—delaying the more fundamental step, namely to recognize and develop a deeper, time-based foundation of physics.